SCIENTISTS: The Lives and Works of 150 Scientists

SCIENTISTS: The Lives and Works of 150 Scientists

Peggy Saari and Stephen Allison, Editors

VOLUME 3

P–Z

U·X·L®
AN IMPRINT OF GALE

Scientists: The Lives and Works of 150 Scientists

Edited by Peggy Saari and Stephen Allison

Staff

Carol DeKane Nagel, *U·X·L Developmental Editor*
Thomas L. Romig, *U·X·L Publisher*

Shanna P. Heilveil, *Production Assistant*
Evi Seoud, *Assistant Production Manager*
Mary Beth Trimper, *Production Director*

Kimberly Smilay, *Permissions Specialist (Pictures)*

Tracey Rowens, *Cover and Page Designer*
Cynthia Baldwin, *Art Director*

Linda Mahoney, *Typesetter*

Library of Congress Cataloging-in-Publication Data

Scientists : the lives and works of 150 physical and social scientists / edited by Peggy Saari and Stephen Allison.
p. cm.
Includes bibliographical references and index.
Contents: v. 1. A-F – v. 2. G-O – v. 3. P-Z
ISBN 0-7876-0959-5 (set); 0-7876-0960-9 (v. 1); 0-7876-0961-7 (v. 2); 0-7876-0962-5 (v. 3)
1. Physical Scientists–Biography. 2. Social scientists–Biography. I. Saari, Peggy. II. Allison, Stephen, 1969– .
Q141.S3717 1996
509'.2'2–dc20
[B] 96-25579
CIP

∞™ This book is printed on acid-free paper that meets the minimum requirements of American National Standard for Information Sciences–Permanence Paper for Printed Library Materials, ANSI Z39.48-1984.

Printed in the United States of America

10 9 8 7 6 5 4

Contents

Albert Einstein

VOLUME 1

VOLUME 2

VOLUME 3

Scientists by Field of Specialization

Italic type indicates volume numbers.

Ruth Patrick

Archaeology

Astronomy and Space

Astrophysics

Atomic/Nuclear Physics

Bacteriology

Biochemistry

Biology

Biomedical Engineering

Biophysics

Botany

Chemistry

Climatology

Computer Science

Environmentalism

Environmental Science

Epidemiology

Ethnobotany

Ethology

Evolutionary Biology

Genetics

Geology

Geophysics

History of Science

Horticulture

Immunology

Invention

Limnology

Logic

Marine Biology

Mathematics

Medicine

Metallurgy

Meteorology

Microbiology

Microscopy

Molecular Biology

Natural History

Neuroendocrinology

Neurology

Oceanography

Organic Chemistry

Paleontology

Pathology

Pharmacology

Philosophy of Science

Physical Chemistry

Physics

Physiology

Psychiatry

Psychology

Radiology

Rocket Science

Seismology

Sociobiology

Surgery

Virology

Zoology

Reader's Guide

An Wang

Budding scientists and those entering the fascinating world of science for fun or study will find inspiration in these three volumes. *Scientists: The Lives and Works of 150 Scientists* presents detailed biographies of the women and men whose theories, discoveries, and inventions have revolutionized science and society. From Louis Pasteur to Bill Gates and Elijah McCoy to Margaret Mead, *Scientists* explores the pioneers and their innovations that students most want to learn about.

Scientists from around the world and from the Industrial Revolution until the present day are featured, in fields such as astronomy, ecology, oceanography, physics, and more.

In *Scientists* students will find:

- 150 scientist biographies, each focusing on the scientist's early life, formative experiences, and inspirations—details that keep students reading
- "Impact" boxes that draw out important information and sum up why each scientist's work is indeed revolutionary

- 120 biographical boxes that highlight individuals who influenced the work of the featured scientist or who conducted similar research
- Sources for further reading so students know where to delve even deeper
- More than 300 black-and-white portraits and additional photographs that give students a better understanding of the people and inventions discussed

Each *Scientists* volume begins with a listing of scientists by field, ranging from aeronautical engineering to zoology; a timeline of major scientific breakthroughs; and a glossary of scientific terms used in the text. Volumes conclude with a cumulative subject index so students can easily find the people, inventions, and theories discussed throughout *Scientists.*

Acknowledgment

Aaron Saari was an important asset to this project. In addition to conducting research, he wrote the biographical boxes and several main entries.

Suggestions

We welcome any comments on this work and suggestions for individuals to feature in future editions of *Scientists.* Please write: Editors, *Scientists,* U•X•L, Gale Research, 835 Penobscot Bldg., Detroit, Michigan 48226-4094; call toll-free: 800-877-4253; or fax to: 313-961-6348.

Timeline of Scientific Breakthroughs

Donald Johanson and Lucy

1730s **Charles Townshend** introduces innovative farming methods that help spur the Industrial Revolution in England.

1769 **Richard Arkwright** patents the water frame, a spinning machine powered by a water wheel.

1769 **James Watt** patents his design for the steam engine.

1774 **Joseph Priestley** reports the results of his experiments with oxygen, gets credit for discovering the gas.

1789 **William Herschel** completes his revolutionary study of the nature of the universe.

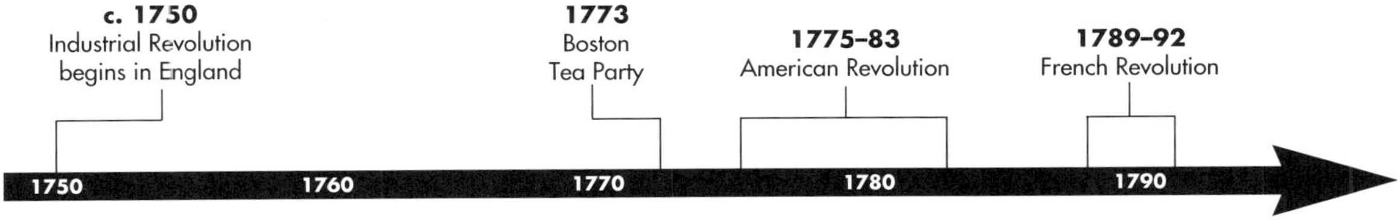

1789 **Antoine Lavoisier** describes the role of oxygen in animal and plant respiration.

1799 **Carl Friedrich Gauss** discovers the root form of all algebraic equations.

1808 **John Dalton** publishes his view of the atomic theory of matter, marking the beginning of modern chemistry.

1808 **Humphry Davy** invents the carbon arc lamp, initiating the entire science of electric lighting.

1823 **Charles Babbage** begins to build his Difference Engine, predecessor of the modern digital computer.

1831 **Michael Faraday** confirms that electricity and magnetism are a single force.

1847 **James Prescott Joule** publishes his calculation for the mechanical equivalent of heat.

1848 **William Thomson, Lord Kelvin** develops the Kelvin scale of absolute temperature.

1854 **Louis Pasteur** begins his experiments with fermentation that lead to the widespread sterilization of foods.

1856 **Henry Bessemer** patents the converter, a device that will revolutionize the steelmaking industry.

1859 **Charles Darwin** publishes *The Origin of Species by Means of Natural Selection,* "the book that shook the world."

1864–73 **James Clerk Maxwell** devises equations that prove that magnetism and electricity are distinctly related.

1865 **Gregor Mendel** presents his basic laws of heredity.

1867 **Alfred Nobel** patents dynamite.

1869 **Dmitry Mendeleev** formulates the periodic law.

1803
United States and France complete the Louisiana Purchase

1829
George Stephenson develops the railroad

1837
Queen Victoria II begins reign over England

1848
Karl Marx and Friedrich Engels write the *Communist Manifesto*

1800 1810 1820 1830 1840 1850

1871 **Luther Burbank** develops the Burbank potato.

1872 **Elijah McCoy** patents the lubricating cup for steam engines that eventually becomes known as "the real McCoy."

1876 **Alexander Graham Bell** patents the telephone.

1877 **Thomas Alva Edison** invents the phonograph; two years later he demonstrates in public the first incandescent lightbulb.

1884 **Svante Arrhenius** formulates electrolytic dissociation theory, explaining how some substances conduct electricity in solutions.

1887 **Nikola Tesla** perfects the use of alternating-current electricity in his polyphase motor.

1887 **Granville T. Woods** patents the railway telegraph, allowing communication between moving trains.

1892 **Rudolf Diesel** patents an internal-combustion engine superior to the gasoline-powered engines of the day.

1895 **Wilhelm Röntgen** discovers X rays.

1896 **George Washington Carver** joins the Tuskegee Institute, beginning his career in scientific agriculture.

1897 **Guglielmo Marconi** patents the wireless radio.

1898 **Marie Curie** and **Pierre Curie** publish the first in a series of papers on radioactivity.

1899 **Sigmund Freud** lays out the basic principles of psychoanalytic theory in *The Interpretation of Dreams.*

1900 **David Hilbert** sets an agenda of twenty-three problems for mathematicians to solve during the twentieth century.

1900 **Max Planck** formulates the quantum theory, taking the science of physics into the modern age.

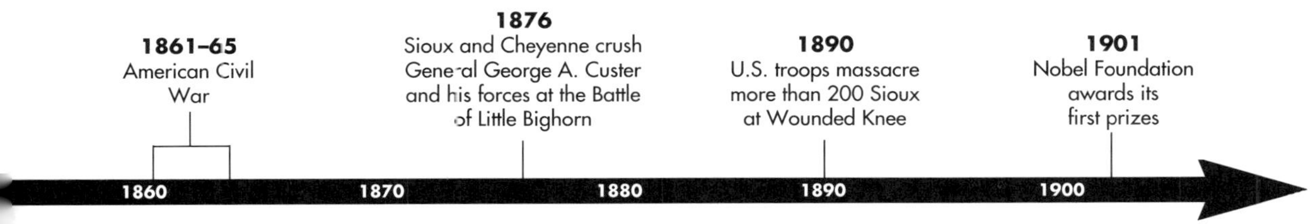

1900 **Florence R. Sabin** commences her studies of the human lymphatic system.

1903 **Bertrand Russell** unites logic and mathematics in *The Principals of Mathematics.*

1903 **Wilbur Wright** and **Orville Wright** achieve the first sustained flights in a power-driven aircraft.

c. 1903 **Arthur C. Parker** conducts his first formal archaeological excavation on the Cattaraugus Reservation.

1905 **Albert Einstein** formulates his special theory of relativity, changing forever the way scientists look at the nature of space, time, and matter.

1906 **Lee De Forest** patents the audion tube (triode), crucial to the development of the modern radio.

1910 **Annie Jump Cannon**'s method for classifying stars, known as the Harvard system, is adopted by the astronomical community.

1910 **Ernest Rutherford** determines the structure of the atom.

1912 **Alfred Wegener** proposes the theory of continental drift.

1913 **Niels Bohr** introduces the quantum mechanical model of the atom.

1913 **C. G. Jung** publishes *The Psychology of the Unconscious* and is outcast by the psychoanalytic community.

1919 **Karl von Frisch** discovers that honeybees communicate through ritual dances.

1919 **Karl Menninger** and his father open the Menninger Clinic for psychiatric treatment and research.

1920–40 **Edith H. Quimby** conducts studies of the effects of radiation on the body.

1924 **Vladimir Zworykin** patents the kinescope, or the picture tube, which will make television as we know it possible.

1926 **Robert H. Goddard** launches the first liquid-propellant rocket.

1926 **Ivan Pavlov** publishes his masterwork, *Conditioned Reflexes.*

1927 **William Augustus Hinton** develops a test that becomes the standard for diagnosing syphilis.

1928 **Alexander Fleming** discovers penicillin.

1928 **Margaret Mead** publishes *Coming of Age in Samoa.*

1929 **Edwin Hubble** initiates the theory of an expanding universe.

1930s–40s **B. F. Skinner** conducts experiments that convince him that behavior can be controlled through the environment.

c. 1932 **Ruth Patrick** begins studying the presence of diatoms in various marine ecosystems.

1935 **Subrahmanyan Chandrasekhar** rocks the astronomical community with his radical theories on the evolution of white dwarf stars.

1935 **Percy L. Julian** synthesizes physostigmine, a chemical used in the treatment of glaucoma.

1935 **Charles F. Richter** develops the Richter scale to measure earthquake intensity.

1935 **Robert Watson-Watt** develops radar.

1929 Great Depression begins

1933 Adolf Hitler and Nazi Party take control of Germany

1936 Spanish Civil War begins

1939 John Steinbeck publishes *The Grapes of Wrath*

1928 1931 1934 1937 1940

c. 1935 **Berta Scharrer** and her husband Ernst begin their pioneering work in neuroendocrinology.

1935–38 **Konrad Lorenz** spends his "goose summers" confirming his many hypotheses on animal behavior patterns.

1938 **Lise Meitner,** with Otto Robert Frisch, develops the theory that explains nuclear fission.

c. 1938 **Katharine Burr Blodgett** invents nonreflecting glass.

1939 **Charles Richard Drew** develops a method to process and preserve blood plasma through dehydration.

1939 **Lloyd A. Hall** cofounds the Institute of Food Technologies.

1939 **Ernest Everett Just** publishes his findings on the role protoplasm plays in a cell.

1939 **Linus Pauling** publishes the landmark *Nature of the Chemical Bond and the Structure of Molecules and Crystals.*

1942 **Enrico Fermi** produces the first self-sustaining nuclear chain reaction.

1942 The World Health Organization adopts **Florence Seibert**'s skin test as the standard tool for diagnosing tuberculosis.

1944 **Norman Borlaug** shares agricultural advances with the Mexican government, thus starting the Green Revolution.

1944 **George H. Hitchings** and **Gertrude Belle Elion** begin their twenty-five-year collaboration developing "rational" drugs.

1939–45
World War II

1946
Cold War between the United States and the Soviet Union begins

1949
Mao Zedong named first leader of People's Republic of China

1950–53
Korean War

1942 1944 1946 1948 1950 1952

1944 **Barbara McClintock** begins her studies that will lead to her discovery of "jumping genes."

1944 **Vivien Thomas** perfects the surgical technique that will save thousands of "blue babies."

1945 **J. Robert Oppenheimer** sees the culmination of his work as director of the Manhattan Project: two atomic bombs are dropped on Japan to end World War II.

1947 **Carl Ferdinand Cori** and **Gerty T. Cori** share the Nobel Prize for their studies on sugar metabolism, begun in the early 1920s.

1947 **Thor Heyerdahl** crosses the Pacific on the balsa raft *Kon-Tiki.*

1947 **Edwin H. Land** introduces the instant camera.

1947 **William Shockley** and his research team develop the transistor.

1948 **Dorothy Hodgkin** begins using X-ray crystallography to determine the structure of vitamin B^{12}.

1948 **Norbert Wiener** publishes *Cybernetics,* detailing his theories on control and communication in humans and machines.

c. 1948 **Albert Baez,** with Paul Kirkpatrick, builds the first X-ray microscope.

1949 **Maria Goeppert-Mayer** publishes her hypothesis for the structure of atomic nuclei.

1951 **R. Buckminster Fuller** patents the geodesic dome.

1951–52 **Jacques Cousteau,** with the *Calypso,* undertakes his first extensive expedition.

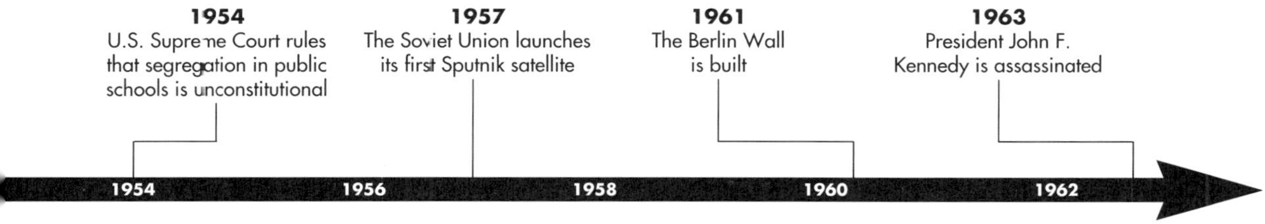

1952 Through X-ray diffraction, **Rosalind Franklin** makes a preliminary determination of the structure of DNA.

1952 **Dorothy Horstmann** concludes that the virus that causes polio travels through the bloodstream to reach the nervous system.

1953 **Auguste Piccard** and **Jacques Piccard** set a depth record of almost two miles in their bathyscaphe *Trieste*.

1953 **James D. Watson** and **Francis Crick** unravel the mystery of the structure of DNA.

1954 **Angeles Alvariño** begins groundbreaking studies of marine zooplankton.

1955 **Jonas Salk**'s polio vaccine is pronounced effective, potent, and safe.

1955 **An Wang** patents techniques for developing magnetic core memories in computers.

1956 **Charlotte Friend** isolates the virus responsible for leukemia in mice, thereby pointing the way for future cancer research.

1956 **Chen Ning Yang** and **Tsung-Dao Lee** theorize that the unusual behavior of K-mesons violates the law of conservation of parity.

c. 1956 **Rosalyn Sussman Yalow** codevelops the diagnostic tool radioimmunoassay (RIA).

1957 **Albert Sabin** begins administering his oral polio vaccine to millions of Russian children.

1957 **Chien-Shiung Wu** conducts beta decay experiments that confirm the violation of the conservation of parity.

1958 **James Van Allen** discovers bands of high-level radiation surrounding Earth.

1965 Malcolm X is assassinated

1966 National Organization for Women (NOW) is founded

1968 Martin Luther King Jr., is assassinated

1969 U.S. astronaut Neil Armstrong walks on the Moon

1971 Supreme Court orders busing to desegregate schools

1964 1966 1968 1970 1971 1972

c. 1958 **Meredith Gourdine** develops the formula for electrogasdynamics.

1959 **Luis Alvarez** develops a 72-inch bubble chamber to better study subatomic particles.

1959 **Mary Leakey** finds the 1.75 million-year-old fossilized skull of a hominid that **Louis S. B. Leakey** names "East Africa man."

1959 **Marvin Minsky** cofounds the Artificial Intelligence Project at the Massachusetts Institute of Technology (MIT).

1959 **Robert Noyce** invents the integrated circuit, or microchip, revolutionizing twentieth-century technology.

1960 **George Bass** undertakes the first excavation of an underwater shipwreck site.

1960 **Jane Goodall** establishes a camp in the Gombe Stream Reserve, beginning her long-term studies of chimpanzee behavior.

1960 **Theodore Maiman** constructs the first working laser.

1960s **Tetsuya Theodore Fujita** develops the F Scale, used to measure the strength of tornadoes.

1962 **Rachel Carson** publishes *Silent Spring,* sparking the beginning of the environmental movement.

1966 **Keiiti Aki** develops the seismic moment, a new method of measuring the magnitude of earthquakes.

1967 **E. Margaret Burbidge** and **Geoffrey Burbidge** publish *Quasi-Stellar Objects,* one of the first surveys of quasars.

1967 **Jocelyn Bell Burnell** discovers pulsars.

1968 **James E. Lovelock** publishes his controversial Gaia theory about Earth's regulation of her ecosystems.

1973
Americans pull out of Vietnam

1974
Richard M. Nixon resigns the presidency of the United States

1976
America celebrates its bicentennial

1977
U.S. president Jimmy Carter unveils policy designed to reduce energy consumption

1973 **1974** **1975** **1976** **1977**

1968 **Abdus Salam** announces his theory of the electroweak force.

1968 **Edward O. Wilson** confirms his theory of species equilibrium in a study of insect life on six islands off the Florida keys.

1970 **Bruce N. Ames** develops test for measuring the cancer-causing potential of chemicals.

1970 **Sylvia A. Earle** spends two weeks in an underwater chamber to study marine habitats.

1970 **Paul R. Ehrlich** helps found Zero Population Growth, a group that aims to educate people on the environmental dangers caused by overpopulation.

1971 **Fred Begay** begins his research into harnessing the clean, safe power of nuclear fusion.

1971 **Helen Caldicott** first organizes opposition to nuclear weapons, prompting the French government to cease atmospheric testing for a time.

1971 **Geoffrey Hounsfield** tests the first CAT scan machine.

1972 **Stephen Jay Gould** introduces the concept of punctuated equilibrium, contradicting a central tenet of the Darwinian theory of evolution.

1972 **Richard Leakey** unearths a 1.9-million-year-old *Homo habilis* skull, the oldest *H. habilis* specimen discovered so far.

1973 **Shirley Ann Jackson** is the first African American woman to earn a doctorate from MIT.

1974 **Paul Berg** publishes the "Berg letter," warning of the dangers of genetic engineering, a field he pioneered in the late 1960s and 1970s.

1977 *Star Wars* is realeased

1979 Three Mile Island nuclear reactor melts down

1981 AIDS is first recognized

1982 Deadline for states to ratify Equal Rights Amendment (ERA) expires

1983 U.S. invades Grenada

1978 · 1979 · 1980 · 1981 · 1982 · 1983

1974 **Donald Johanson** finds Lucy, the oldest fossilized remains of a hominid ever unearthed.

c. 1974 **Stephen Hawking** discovers that black holes emit radiation.

1975 **Sandra Faber,** with Robert Jackson, formulates the first galactic sealing law, used to help calculate distances between galaxies.

1975 **Bill Gates** begins his career as a software designer and entrepreneur when he cofounds Microsoft.

1976 **Maxine Singer** helps formulate guidelines for responsible biochemical genetics research.

c. 1976 **James S. Williamson** supervises the design of the first solar-powered electricity-generating plant in the United States.

1977 **Steven Jobs** and Stephen Wozniak introduce the Apple II computer, touching off the personal computer revolution.

1977 **Louis Keith** and his twin brother found the Center for the Study of Multiple Births.

1978 **Elizabeth H. Blackburn** begins groundbreaking work that leads to the discovery of the enzyme telomerase.

1978 **Patrick Steptoe** produces the first test-tube baby.

1980 **Lynn Conway** publishes *Introduction to VLSI Systems,* which simplifies the way computer chips are produced.

1980 **Levi Watkins Jr.,** performs the first implantation of the automatic heart defibrillator.

1982 **Robert K. Jarvik** implants the Jarvik-7 artificial heart into Barney Clark.

1984
Famine and drought in Africa kills 300,000 people

1986
Space shuttle *Challenger* explodes

1988
Toni Morrison becomes the first African American to win Nobel Prize for literature

1984 1985 1986 1987 1988

1983 **Irene Diggs** publishes *Black Chronology,* focusing on the accomplishments of people of African descent.

1984 **Helene D. Gayle** begins research on the effects of AIDS on children throughout the world.

1984 **Jaron Lanier** founds VPL Research Inc., to provide virtual reality software to the general computer user.

1985 Using technology he calls telepresence, **Robert D. Ballard** discovers the remains of the sunken *Titanic.*

1986 **Sally Fox** opens a mail-order business to sell the naturally colored cotton fibers she developed.

1986 **Margaret Geller** reports finding a "Great Wall" of galaxies, forcing some cosmologists to rethink existing theories of the beginning of the universe.

c. 1986 **Francisco Dallmeier** begins biodiversity research and education programs in several countries.

1987 **Anthony S. Fauci**'s research team discovers how the AIDS virus is transmitted.

1987 **Oliver Wolf Sacks**'s 1973 book *Awakenings,* about his treatment of patients with sleeping sickness, is made into a popular film.

1990 **Walter Gilbert** starts the human genome project to compile a genetic map of the entire human being.

1993 **Mark Plotkin** publishes *Tales of a Shaman's Apprentice,* discussing his excursions into the Amazon rain forest in search of medicinal plants and traditions.

1989 Berlin Wall is destroyed

1990–91 Persian Gulf War

1992 Los Angeles riots

1993 Apartheid is outlawed in South Africa

1989 1990 1991 1992 1993

Words to Know

Meredith Gourdine

A

Absolute zero: the theoretical point at which a substance has no heat and motion ceases; equivalent to -276°C or -459.67°F.

Algae: a diverse group of plant or plantlike organisms that grow mainly in water.

Alpha particle: a positively charged nuclear particle that consists of two protons and two electrons; it is ejected at a high speed from disintegrating radioactive materials.

Alternating current: the flow of electrons first in one direction and then in the other at regular intervals.

Amino acids: organic acids that are the chief components of proteins.

Anatomy: the study of the structure and form of biological organisms.

Anthropology: the science that deals with the study of human beings, especially their origin, development, divisions, and customs.

Archaeology: the scientific study of material remains, such as fossils and relics, of past societies.

Artificial intelligence: the branch of science concerned with the development of machines having the ability to perform tasks normally thought to require human intelligence, such as problem solving, discriminating among single objects, and response to spoken commands.

Asteroid: one of thousands of small planets located in a belt between the orbits of Mars and Jupiter.

Astronomy: the study of the physical and chemical properties of objects and matter outside Earth's atmosphere.

Astrophysics: the branch of physics involving the study of the physical and chemical nature of celestial objects and events.

Atomic bomb: a weapon of mass destruction that derives its explosive energy from nuclear fission.

B

Bacteria: a large, diverse group of mostly single-celled organisms that play a key role in the decay of organic matter and the cycling of nutrients.

Bacteriology: the scientific study of bacteria, their characteristics, and their activities as related to medicine, industry, and agriculture.

Bacteriophage: a virus that infects bacteria.

Behaviorism: the school of psychology that holds that human and animal behavior is based not on independent will nor motivation but rather on response to reward and punishment.

Beta decay: process by which a neutron in an atomic nucleus breaks apart into a proton and an electron.

Big bang: in astronomy, the theory that the universe resulted from a cosmic explosion that occurred billions of years ago and then expanded over time.

Biochemistry: the study of chemical compounds and processes occurring in living organisms.

Biodiversity: the number of different species of plants and animals in a specified region.

Biology: the scientific study of living organisms.

Biophysics: the branch of biology in which the methods and principles of physics are applied to the study of living things.

Biotechnology: use of biological organisms, systems, or processes to make or modify products.

Botany: the branch of biology involving the study of plant life.

C

Carcinogen: a cancer-causing agent, such as a chemical or a virus.

Cathode: a negatively charged electrode.

Cathode rays: electrons emitted by a cathode when heated.

Chemistry: the science of the nature, composition, and properties of material substances and their transformations.

Chromosome: threadlike structure in the nucleus of a cell that carries thousands of genes.

Circuit: the complete path of an electric current including the source of electric energy; an assemblage of electronic elements.

Climatology: the scientific study of climates and their phenomena.

Combustion: a rapid chemical process that produces heat and light.

Conductor: a substance able to carry an electrical current.

Conservation biology: the branch of biology that involves conserving rapidly vanishing wild animals, plants, and places.

Conservation laws: laws of physics that state that a particular property, mass, energy, momentum, or electrical charge is not lost during any change.

Cosmic rays: charged particles, mainly the nuclei of hydrogen and other atoms, that bombard Earth's upper atmosphere at velocities close to that of light.

Cosmology: the study of the structure and evolution of the universe.

Cross-fertilization: a method of fertilization in which the gametes (mature male or female cells) are produced by separate individuals or sometimes by individuals of different kinds.

Cryogenics: the branch of physics that involves the production and effects of very low temperatures.

Crystallography: the science that deals with the forms and structures of crystals.

Cytology: the branch of biology concerned with the study of cells.

D

Diffraction: the spreading and bending of light waves as they pass through a hole or slit.

Direct current: a regular flow of electrons, always in the same direction.

DNA (deoxyribonucleic acid): a long molecule composed of two chains of nucleotides (organic chemicals) that contain the genetic information carried from one generation to another.

E

Earthquake: an unpredictable event in which masses of rock shift below Earth's surface, releasing enormous amounts of energy and sending out shockwaves that sometimes cause the ground to shake dramatically.

Ecology: the branch of science dealing with the interrelationship of organisms and their environments.

Ecosystem: community of plants and animals and the physical environment with which they interact.

Electrochemistry: the branch of physical chemistry involving the relation of electricity to chemical changes.

Electrodes: conductors used to establish electrical contact with a nonmetallic part of a circuit.

Electromagnetism: the study of electric and magnetic fields and their interaction with electric charges and currents.

Electron: a negatively charged particle that orbits the nucleus of an atom.

Entomology: the branch of zoology dealing with the study of insects.

Environmentalism: the movement to preserve and improve the natural environment, and particularly to control pollution.

Enzyme: any of numerous complex proteins that are produced by living cells and spark specific biochemical reactions.

Epidemiology: the study of the causes, distribution, and control of disease in populations.

Ethnobotany: the plant lore of a race of people.

Ethnology: science that deals with the division of human beings into races and their origin, distribution, relations, and characteristics.

Ethology: the scientific and objective study of the behavior of animals in the wild rather than in captivity.

Evolution: in the struggle for survival, the process by which successive generations of a species pass on to their offspring the characteristics that enable the species to survive.

Extinction: the total disappearance of a species or the disappearance of a species from a given area.

F

Fossils: the remains, traces, or impressions of living organisms that inhabited Earth more than ten thousand years ago.

G

Gamma rays: short electromagnetic wavelengths that come from the nuclei of atoms during radioactive decay.

Gene: in classical genetics, a unit of hereditary information that is carried on chromosomes and determines observable characteristics; in molecular genetics, a special sequence of DNA or RNA located on the chromosome.

Genetic code: the means by which genetic information is translated into the chromosomes that make up living organisms.

Genetics: the study of inheritance in living organisms.

Genome: genetic material of a human being; the complete genetic structure of a species.

Geochemistry: the study of the chemistry of Earth (and other planets).

Geology: the study of the origin, history, and structure of Earth.

Geophysics: the physics of Earth, including studies of the atmosphere, earthquakes, volcanism, and oceans.

Global warming: the rise in Earth's temperature that is attributed to the buildup of carbon dioxide and other pollutants in the atmosphere.

Greenhouse effect: warming of Earth's atmosphere due to the absorption of heat by molecules of water vapor, carbon dioxide, methane, ozone, nitrous oxide, and chlorofluorocarbons.

H

Herpetology: the branch of zoology that deals with reptiles and amphibians.

Hominids: humanlike creatures.

Hormones: chemical messengers produced in living organisms that play significant roles in the body, such as affecting growth, metabolism, and digestion.

Horticulture: the science of growing fruits, vegetables, and ornamental plants.

Hybridization: cross-pollination of plants of different varieties to produce seed.

I

Immunology: the branch of medicine concerned with the body's ability to protect itself from disease.

Imprinting: the rapid learning process that takes place early in the life of a social animal and establishes a behavioral pattern, such as a recognition of and attraction to its own kind or a substitute.

In vitro fertilization: fertilization of eggs outside of the body.

Infrared radiation: electromagnetic rays released by hot objects; also known as a heat radiation.

Infertility: the inability to produce offspring for any reason.

Invertebrates: animals lacking a spinal column.

Ion: an atom or groups of atoms that carries an electrical charge-either positive or negative-as a result of losing or gaining one or more electrons.

Isotopes: atoms of a chemical element that contain the same number of protons but a different number of neutrons.

L

Laser: acronym for light amplification by stimulated emission of radiation; a device that produces intense light with a precisely defined wavelength.

Light-year: in astronomy, the distance light travels in one year, about six trillion miles.

Limnology: the branch of biology concerning freshwater plants.

Logic: the science of the formal principles of reasoning.

M

Magnetic field: the space around an electric current or a magnet in which a magnetic force can be observed.

Maser: acronym for microwave amplification of stimulated emission of radiation; a device that produces radiation in short wavelengths.

Metabolism: the process by which living cells break down organic compounds to produce energy.

Metallurgy: the science and technology of metals.

Meteorology: the science that deals with the atmosphere and its phenomena and with weather and weather forecasting.

Microbiology: branch of biology dealing with microscopic forms of life.

Microwaves: electromagnetic radiation waves between one millimeter and one centimeter in length.

Molecular biology: the study of the structure and function of molecules that make up living organisms.

Molecule: the smallest particle of a substance that retains all the properties of the substance and is composed of one or more atoms.

Mutation: any permanent change in hereditary material, involving either a physical change in chromosome relations or a biochemical change in genes.

N

Natural selection: the natural process by which groups best adjusted to their environment survive and reproduce, thereby passing on to their offspring genetic qualities best suited to that environment.

Nervous system: the bodily system that in vertebrates is made up of the brain and spinal cord, nerves, ganglia, and other organs and that receives and interprets stimuli and transmits impulses to targeted organs.

Neurology: the scientific study of the nervous system, especially its structure, functions, and abnormalities.

Neurosecretion: the process of producing a secretion by nerve cells.

Neurosis: any emotional or mental disorder that affects only part of the personality, such as anxiety or mild depression, as a result of stress.

Neutron: an uncharged particle found in atomic nuclei.

Neutron star: a hypothetical dense celestial object that consists primarily of closely packed neutrons that results from the collapse of a much larger celestial body.

Nova: a star that suddenly increases in light output and then fades away to its former obscure state within a few months or years.

Nuclear fallout: the drifting of radioactive particles into the atmosphere as the result of nuclear explosions.

Nuclear fission: the process in which an atomic nucleus is split, resulting in the release of large amounts of energy.

O

Oceanography: the science that deals with the study of oceans and seas.

Optics: the study of light and vision.

Organic: of, relating to, or arising in a bodily organ

Ozone layer: the atmospheric layer of approximately twenty to thirty miles above Earth's surface that protects the lower atmosphere from harmful solar radiation.

P

Paleoanthropology: the branch of anthropology dealing with the study of mammal fossils.

Paleontology: the study of the life of past geological periods as known from fossil remains.

Particle physics: the branch of physics concerned with the study of the constitution, properties, and interactions of elementary particles.

Particles: the smallest building blocks of energy and matter.

Pathology: the study of the essential nature of diseases, especially the structural and functional changes produced by them.

Periodic table: a table of the elements in order of atomic number, arranged in rows and columns to show periodic similarities and trends in physical and chemical properties.

Pharmacology: the science dealing with the properties, reactions, and therapeutic values of drugs.

Physics: the science that explores the physical properties and composition of objects and the forces that affect them.

Physiology: the branch of biology that deals with the functions and actions of life or of living matter, such as organs, tissues, and cells.

Plankton: floating animal and plant life.

Plasma physics: the branch of physics involving the study of electrically charged, extremely hot gases.

Primate: any order of mammals composed of humans, apes, or monkeys.

Protein: large molecules found in all living organisms that are essential to the structure and functioning of all living cells.

Proton: a positively charged particle found in atomic nuclei.

Psychiatry: the branch of medicine that deals with mental, emotional, and behavioral disorders.

Psychoanalysis: the method of analyzing psychic phenomenon and treating emotional disorders that involves treatment sessions during which the patient is encouraged to talk freely about personal experiences, especially about early childhood and dreams.

Psychology: the study of human and animal behavior.

Psychotic: a person with severe emotional or mental disorders that cause a loss of contact with reality.

Q

Quantum: any of the very small increments or parcels into which many forms of energy are subdivided.

Quasar: celestial object more distant than stars that emits excessive amounts of radiation.

R

Radar: acronym for radio detection and ranging; the process of using radio waves to detect objects.

Radiation: energy emitted in the form of waves or particles.

Radio waves: electromagnetic radiation.

Radioactive fallout: the radioactive particles resulting from a nuclear explosion.

Radioactivity: the property possessed by some elements (as uranium) or isotopes (as carbon 14) of spontaneously emitting energetic particles (as electrons or alpha particles) by disintegration of their atomic nuclei.

Radiology: the branch of medicine that uses X rays and radium (an intensely radioactive metallic element) to diagnose and treat disease.

Redshift: the increase in the wavelength of all light received from a celestial object (or wave source), usually because the object is moving away from the observer.

RNA (ribonucleic acid): any of various nucleic acids that are associated with the control of cellular chemical activities.

S

Scientific method: collecting evidence meticulously and theorizing from it.

Seismograph: a device that records vibrations of the ground and within Earth.

Seismology: the study and measurement of earthquakes.

Semiconductor: substances whose ability to carry electrical current is lower than that of a conductor (like metal) and higher than that of insulators (like rubber).

Shortwave: a radio wave having a wavelength between ten and one hundred meters.

Sociobiology: the systematic study of the biological basis for all social behavior.

Solid state: using semiconductor devices rather than electron tubes.

Spectrum: the range of colors produced by individual elements within a light source.

Steady-state theory: a theory that proposes that the universe has neither a beginning nor an end.

Stellar spectra: the distinctive mix of radiation emitted by every star.

Stellar spectroscopy: the process that breaks a star's light into component colors so that the various elements of the star can be observed.

Sterilization: boiling or heating of instruments and food to prevent proliferation of microorganisms.

Supernova: a catastrophic explosion in which a large portion of a star's mass is blown out into space, or the star is entirely destroyed.

T

Theorem: in mathematics, a formula, proposition, or statement.

Thermodynamics: the branch of physics that deals with the mechanical action or relations of heat.

Trace element: a chemical element present in minute quantities.

Transistor: a solid-state electronic device that is used to control the flow of electricity in electronic equipment and consists of a small block of semiconductor with at least three electrodes.

V

Vaccine: a preparation administered to increase immunity to polio.

Vacuum tube: an electric tube from which all matter has been removed.

Variable stars: stars whose light output varies because of internal fluctuations or because they are eclipsed by another star.

Variation: in genetics, differences in traits of a particular species.

Vertebrate: an animal that has a spinal column.

Virology: the study of viruses.

Virtual reality: an artificial computer-created environment that seeks to mimic reality.

Virus: a microscopic agent of infection.

W

Wavelength: the distance between one peak of a wave of light, heat, or energy and the next corresponding peak.

X

X ray: a form of electromagnetic radiation with an extremely short wavelength that is produced by bombarding a metallic target with electrons in a vacuum.

Z

Zoology: the branch of biology concerned with the study of animal life.

Zooplankton: small drifting animal life in the ocean.

SCIENTISTS: The Lives and Works of 150 Scientists

Arthur C. Parker

Born in 1881
Cattaraugus Indian Reservation, New York
Died January 1, 1955

Native American anthropologist and social activist Arthur C. Parker systematically collected Seneca folklore and oral history.

One of the most prominent Native American intellectuals of the first half of the twentieth century, Arthur C. Parker made significant contributions to the fields of anthropology, social activism, and museology. Aware that traditional Native American cultures were rapidly changing, impacted by the mainstream American culture that was also undergoing great change, Parker believed that Native Americans would be able to adjust to a new American society and maintain their cultural identities in the process.

Learns about his people

Arthur Caswell Parker was born on the Cattaraugus Indian Reservation in New York in 1881, the son of Frederick Ely Parker and Geneva H. Griswold. His father was a half-blood Seneca who worked for the New York Central Railroad, and his mother was a white teacher who had worked on both the Cattaraugus and the Allegany Reservations. Some mem-

> **IMPACT**
>
> Although Seneca anthropologist Arthur C. Parker never obtained a degree from a university, his many contributions to preserving Seneca culture earned him the respect of his colleagues and the Seneca people. Parker systematically collected the folklore and oral history of the Seneca, and he established a museum program through which Iroquois artisans could create and sell crafts, thereby providing jobs as well as a way to maintain cultural traditions. Parker noted that although Native American culture and traditions were sometimes compromised in attempts to adjust to modern society, he believed that Native Americans could adapt to mainstream American culture without losing their identities.

bers of Parker's family were missionaries to the Seneca Indians, and the family was one of the most prominent on the reservation. Parker's grandfather had attended the State Normal School in Albany and was a prosperous farmer. He had also held the office of chief clerk of the Seneca Nation of Indians for many years. A role model for Parker during his youth, he read the classics to his grandson and acquainted him with the Seneca Indians' long and turbulent history.

Adopted into Bear Clan

Because of matrilineal (determined by the mother) rule for legally determining tribal and clan affiliation among the New York Iroquois, neither Parker nor his father were legally recognized as Seneca (the Seneca are one of six nations that banded together to form the Iroquois Confederacy). However, they were both adopted into the Bear Clan, and in his early twenties Parker was given the adult name Gawasowanah (Big Snowsnake). He was also initiated into the Little Water Society, one of the so-called secret medicinal societies of the Iroquois. Parker's acceptance into the group was a particular honor, for many of the members practiced native religions while Parker himself was a recognized Christian. As a result of the traditional rites, Parker considered himself to be a Seneca by birth, culture, and affiliation.

Parker spent most of his childhood on the Cattaraugus Reservation. In 1892 his father was transferred to White Plains, New York, but the family made frequent trips back to the reservation. Parker and his sister attended public school, and he graduated from high school in White Plains in 1897. For a brief period in 1899 he was a student at Centenary Collegiate Institute in Hackettstown, New Jersey, but apparently left school after a fire. Later that year Parker went on to study for the ministry at Williamsport Dickinson Seminary in Pennsylvania. He left Dickinson in 1903 without graduating. During this time Parker worked for a while as a newspaper reporter, developing the facility to write well, an ability that he demonstrated in his scientific writing.

Becomes interested in anthropology

While at the seminary Parker also attended informal classes at the American Museum of Natural History in New York City, conducted by the early anthropologist Frederick Ward Putnam, and even accepted a brief appointment as an archaeological assistant at the museum in 1900. Parker became a frequent visitor at the salon of Harriet Maxwell Converse, an amateur scholar of Iroquois culture who had also been adopted by the Seneca. It was at "Aunt Hattie's" salon that Parker made the acquaintances, both Native American and Anglo-American, who were to become important figures in his life.

In New York Parker also became acquainted with the famous anthropologist Franz Boas, who taught at Columbia College. Parker ultimately rejected Boas's urging to pursue a degree in anthropology. (Anthropologists study humankind in relation to classification and relationships of race, physical character, environment, and culture.) This was a professional decision Parker later regretted, but his reluctance may have been due, in part, to Boas's rejection of Lewis Henry Morgan's (1818–1881) evolutionary theories. Parker felt he owed a particular debt to Morgan, a New York anthropologist who became interested in and studied the local Native Americans and was adopted into the Seneca tribe as Parker had been. Morgan believed that cultures of the world could be classified into progressive stages: savagery, barbarism, and civilization.

Parker took a position as a field archaeologist with Mark R. Harrington, under the direction of Putnam and the Peabody Museum at Harvard University. He conducted his first formal archaeological excavation on the Cattaraugus Reservation, where he systematically collected the folklore (traditional myths and stories) and oral history (in the form of interviews) of the Seneca. In 1904 he took a temporary job at the New York State Library and Museum to collect ethnographic (cultural) materials and art. Also that year Parker married his first wife, Beatrice Tahamont, an Abenaki Indian.

Begins career

In 1905 Parker successfully completed the civil service examination for the position of archaeologist for the New York State Museum. Thus began an important and prolific anthropological career, during which he published extensively on numerous subjects, including archaeology, ethnology (science that deals with the division of human beings into races), folklore, race relations, and museology (the science of museum organization and management). After nine years Parker left the State Museum to become director of the Rochester Museum (now the Rochester Museum and Science Center). Parker continued his professional affiliation with the Rochester Museum until his death in 1955.

Advocates Pan-Indianism

In addition to his career in anthropology, Parker worked vigorously to see Native Americans integrated into (brought into and seen as equals) the fabric of mainstream American society. In 1911 Parker joined the newly forming Society of American Indians (SAI). This organization, which would become the leading Native American advocacy (political support) group of its day, promoted the idea of Pan-Indianism, or greater cooperation among Native Americans. Parker and his associates in SAI grappled with such issues as the effect of mainstream integration on the traditional cultures of Indian tribes. They were philosophically committed to the concept that the American Indian was not vanishing as the result of cultural change, and that Indians could and would adapt to contemporary American culture without losing their identity. Ultimately Parker embraced the melting-pot theory regarding the future of the Native American, both as individuals and as a race, in

Archie Phinney, Nez Percé Anthropologist and Ethnographer

Native American anthropologist, ethnographer (cultural researcher), and activist Archie Phinney (1903–1949) is credited with preserving the traditional language and folklore of the Nez Percé culture through his scholarly work. Before his untimely death at the age of forty-six, Phinney, a member of the Nez Percé nation, authored two anthropology books as well as several journal articles. Phinney's work was internationally recognized; indeed, he was awarded an honorary degree from the Academy of Sciences in Leningrad. For his efforts as both a scholar and Nez Percé community leader—he held positions of leadership with the Bureau of Indian Affairs and the National Congress of American Indians—Phinney was presented with the prestigious Indian Council Fire Award in 1946.

which Native Americans and Anglo-Americans would blend socially and culturally.

Through his involvement with the SAI, Parker became the founder and editor of the *Quarterly Journal,* the official publication of the organization. He eventually became secretary-treasurer and later president of the society as well. Parker resigned from the SAI by 1920, having become the president of the New York Welfare Society.

Stays politically involved

During the Great Depression in the 1930s and continuing up to the start of World War II, Parker obtained federal money for the Rochester Museum to fund an Iroquois arts and crafts program. By paying wages to Iroquois craftsmen and artists for producing a wide variety of traditional items, Parker helped maintain and continue the Seneca culture. This program was also instrumental in providing the material that is the basis of the outstanding Iroquois collection at the museum. In his later years Parker maintained his role as an activist. He founded the Philosophical Society of Albany and the New York State Archaeological Association. In 1944 he attended the inaugural meeting of the Congress of American Indians (NCAI) in Denver, Colorado.

Parker's leadership in Indian reform has directly influenced subsequent movements up to the present day despite the fact that he was faced with attempting to reconcile major cultural and societal questions that remain unsettled to this day. And although Parker never earned a college degree, several universities recognized his contributions in the field of anthropology and museology. He was awarded an honorary master's degree from the University of Rochester in 1922, an honorary doctorate in science from Union College in 1940, and an honorary doctorate of human laws from Keuka College in 1945. He published more than three hundred papers. His interests ranged from archaeology, ethnobotany (the plant lore of a race or people), social and political organizations, history, museum studies, biography, contemporary art, and social activism. Parker died suddenly of a heart attack on New Year's Day, 1955. He was survived by his second wife, two daughters, and a son.

Further Reading

Hertzberg, Hazel W., *The Search for an American Indian Identity: Modern Pan-Indian Movements,* Syracuse University Press, 1971.

Liberty, Margot, ed., "Arthur C. Parker—Seneca, 1881–1955," *American Indian Intellectuals, 1976 Proceedings of the American Ethnological Society,* West Publishing Co., 1978, pp. 128–38.

Parker, Arthur C., *Seneca Myths and Folk Tales,* Buffalo Historical Society, 1923, reprinted, University of Nebraska Press, 1989.

Louis Pasteur

Born December 27, 1822
Dôle, France
Died September 28, 1895

Celebrated French chemist and microbiologist Louis Pasteur saved the failing French wine industry when he discovered what was causing wine to spoil. He rescued the silk industry as well, when he determined what was destroying the silkworms and their food source. Pasteur consequently applied his findings about wine-making and fermentation to milk and similar liquids, and "pasteurization" was introduced. Pasteur also determined that bacteria, along with other germs called viruses, cause infectious diseases, and in his final years he developed the first vaccines to protect animals and people against many fatal diseases.

Louis Pasteur made countless contributions to science, most notably vaccines for such deadly infections as rabies, cholera, and anthrax.

Early life

Louis Pasteur was born on December 27, 1822, in the eastern French town of Dôle, not far from the city of Dijon. His father was a tanner, a craftsman who prepares leather for use in clothes, shoes, and equipment. Pasteur was the only boy in the family, but he had several sisters with whom he was very close. When he was still quite young his family moved to nearby Arbois, a town larger than Dôle. Pasteur grew up and attended elementary school there. Although he was only an average student, he always had the idea that he would be a professor. His family encouraged his ambitious streak; becoming a professor would be a fine achievement for a tanner's son from a small country town.

Pasteur's strongest passion as a boy was painting and drawing. He practiced hard and had talent, producing detailed and accurate portraits of his family and friends. When he was sixteen he briefly considered becoming a painter—and perhaps a professor of art. But he found another interest during his college years. At nineteen he earned his bachelor of science degree and put down his drawing pen and paint brush forever. He had decided to be a scientist.

IMPACT

Louis Pasteur, who developed vaccines against such diseases as cholera and anthrax, founded the Pasteur Institute in 1888, originally to promote research on rabies. He remained the director until his death in 1895. The institute has since grown into a world center for research on infectious diseases, with thousands of employees at sites throughout France. It leads the way in acquired immunodeficiency syndrome (AIDS) research, for example, sharing credit for the discovery of the virus believed to cause the disease. Many Pasteur Institute scientists have won Nobel Prizes for their work.

Studies crystals and light

In 1841 Pasteur applied for the graduate studies program in science at the École Normale Supérieure, the famous school for teachers in Paris. Coming in sixteenth in the entrance exams, he was accepted, but sixteenth was not a good enough result for the demanding Pasteur. He went back and studied on his own for a year. After placing fifth the following year, he agreed to enter graduate school and in 1847 earned his doctor of philosophy degree.

Pasteur's teachers and friends at the École had been working on a new area of study: the makeup of crystals (clear colorless glasses of superior quality) and how they affect light that passes through them. Pasteur devoted his doctoral research to the examination of crystals and light, concentrating on crystals found suspended in tartaric acid and racemic acid.

(Tartaric acid was formed when making wine; racemic acid was formed only rarely during wine making.)

In general, acids are substances that release hydrogen during chemical reactions. Pasteur's work on tartaric and racemic acids led to his first major discovery and gave an impressive boost to his career. Tartaric and racemic acids had the same chemical composition and structure, differing only in how they affected polarized light. (Normally light waves vibrate in many directions. Polarized light waves vibrate in one direction only.) Tartaric acid changed the direction of polarized light that passed through its crystals while racemic acid did not. Pasteur wondered how this could be if they were chemically identical. With a series of careful experiments, Pasteur showed that racemic acid actually had two kinds of crystals: one was identical to those in tartaric acid; the other was the mirror image of the first. Both crystals contained the same atoms, but those atoms were arranged in completely opposite ways. This difference accounted for the different ways the crystals affected light. Pasteur's discovery of these mirror-image molecules immediately established his scientific reputation.

Becomes a professor

In 1848, soon after his first discovery, Pasteur fulfilled his childhood dream by becoming a professor of chemistry at the University of Strasbourg. The following year he married a young woman named Marie Laurent, the daughter of a university official. The couple would eventually have five children, but only two survived into adulthood: a son, Jean-Baptiste, and a daughter, Marie-Louise. Pasteur continued his work on mirror-image molecules, making further breakthroughs. He discovered that usually only naturally occurring organic molecules (molecules produced by living organisms) changed the direction of polarized light. Tartaric acid was one such compound.

In 1854 the Pasteurs moved to Lille in northern France, where Pasteur had been offered a teaching post at the univer-

Joseph Lister, English Surgeon

English surgeon Joseph Lister (1827–1912), drawing on the work of Louis Pasteur, developed antiseptic (sterile) surgery, saving innumerable patients from the dreadful pain and death of postsurgical infection. Lister had become increasingly disturbed by the high rate of often fatal infection that developed in his patients after surgery. After Thomas Anderson, a professor of chemistry and friend of Lister, drew Lister's attention to Pasteur's ideas, Lister immediately concluded that the microorganisms described by Pasteur could also carry wound infections through the air. He developed a method to destroy these organisms using carbolic acid (also known as phenol), a poisonous compound that, when diluted, can be used as a disinfectant.

Lister first used his new antiseptic surgical technique in 1865. Although this and many subsequent operations proved the effectiveness of Lister's antiseptic method, his ideas were vigorously opposed by many of his fellow physicians, who thought the procedures were ridiculously complicated and irrelevant. The significance of Lister's innovations were finally recognized in the late 1870s and 1880s.

sity. Shortly thereafter a local factory owner named Bigo asked for Pasteur's assistance with the fermentation process, the process by which a substance undergoes chemical change by use of a catalyst, such as yeast. Bigo's factory used fermentation to produce alcohol from beet juice, and the owner was hopeful that the latest experiment with tartaric acid could help him preserve his stock. This chance inquiry led to Pasteur's first interest in fermentation and thus to his first steps toward germ theory.

Experiments with fermentation

Pasteur began his research on beet juice by putting samples of the fermenting liquid under a microscope. Like others who had done so, he saw small bits of yeast floating in it. At the time scientists believed that sugar molecules split themselves into simpler alcohol molecules during fermentation. Pasteur's first clue to a different view of fermentation came when he found that the alcohol formed from beet juice changed the direction of polarized light. According to his earlier conclusions, that meant that it was organic. Something in the liquid was alive.

Other scientists who had identified yeast in fermenting liquids considered it a catalyst. After much observation Pasteur agreed that yeast caused fermentation, but careful experiments persuaded him that although it acted as a catalyst, it actually was not one. Yeast, he concluded, was a living organism. (Yeast is actually a single-celled fungus that behaves somewhat like bacteria.) It consumed the beet sugar, leaving alcohol as a waste product. Pasteur went on to demonstrate that other microorganisms, bacteria in this case, cause the juice to go sour by fermenting sugar into lactic acid.

Applying his findings to the French wine industry, which had been plagued with massive spoilage, Pasteur suggested that heating wine gently at about 120°F would kill the bacteria that produced lactic acid and let the wine age properly. Some winemakers were aghast at the thought of introducing heat into the wine-making process, but adding heat did solve the industry's problem.

Focuses on microbiology

In 1857, while pursuing his experiments with fermentation, Pasteur was named director of Scientific Studies at the École Normale Supérieure in Paris. Soon afterward he published his revolutionary findings in a short scientific paper. This work is said to have given birth to the field of microbiology because it marked the first time a real understanding of microscopic organisms had been achieved through experimen-

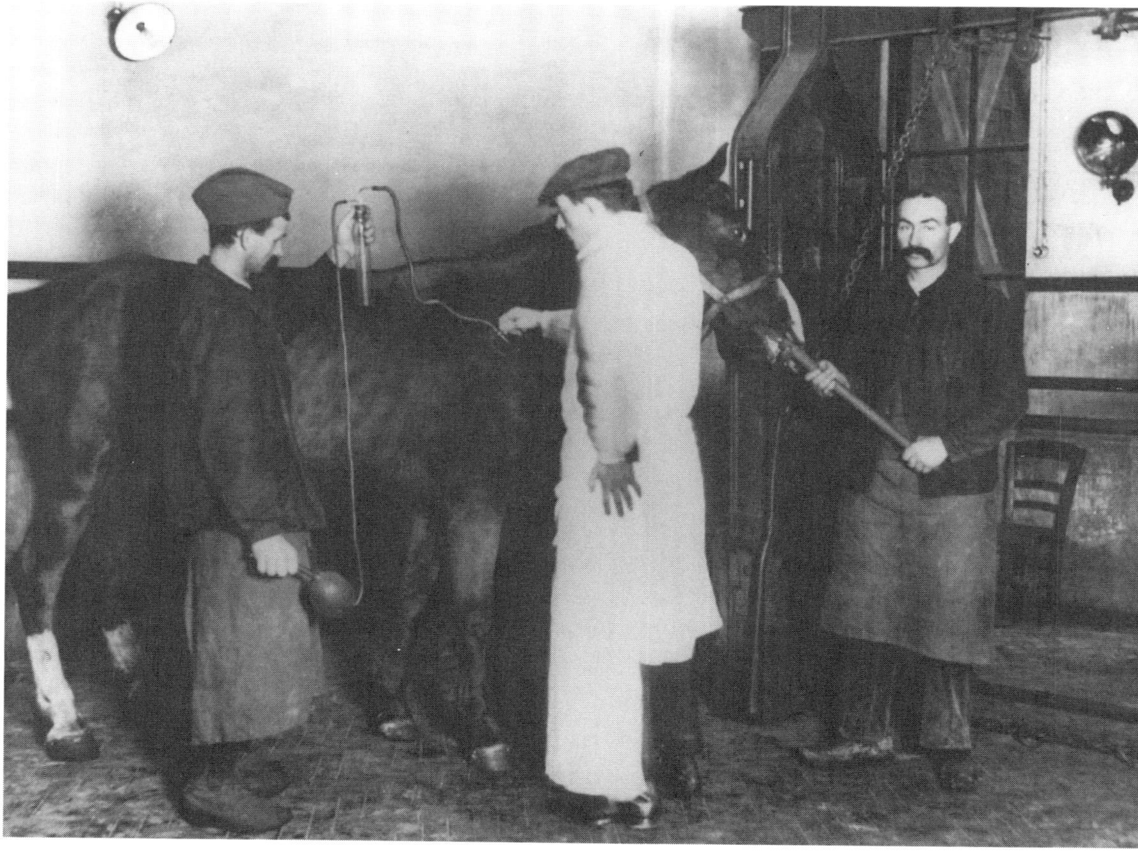

A horse is inoculated in the course of preparing anti-diphtheria serum at the Pasteur Institute.

tation. In the same paper Pasteur mentioned the possibility that his ideas about fermentation might somehow be applied to the problem of infectious diseases (diseases that can be spread).

Solves problems with silkworms

In 1862 representatives of the silk industry approached Pasteur to ask for his assistance with a problem. The silkworms on which the industry relied had been attacked by a mysterious disease; its rapid spread threatened the whole of French silk production. Pasteur accepted the challenge, though he knew nothing about silkworms. For five years he worked on silkworm diseases along with other experiments. He eventually found what was at the root of the problem: a microscopic parasite (an organism living within another organism) that was killing the silkworms and affecting their food, mulberry leaves. His solution seemed quite drastic at the time, but it too worked: destroying all the unhealthy worms and starting with new cultures. The silk industry, very important to the French economy at the time, was saved, and Pasteur's reputation grew.

Disproves spontaneous generation theory

Finding that his job at the École Normale left him too little time to do his experiments, Pasteur resigned in 1867. Napoleon III, the French emperor, agreed to help establish a special laboratory for him at the École. The following year Pasteur suffered a stroke that nearly killed him, leaving his left arm and leg permanently paralyzed. From then on his wife assisted him in his experiments. A few years earlier his father and his two-year-old daughter, Camille, had died; shortly after that, another daughter, Cecile, age twelve, died of typhoid fever. Yet it was during this time of deep suffering that Pasteur accomplished some of his greatest work.

Until the end of the nineteenth century some scientists believed that bacteria arose by themselves, most likely in dead animals or plants. (This theory was called spontaneous generation.) Others disagreed, creating a major controversy in the scientific world. During his early experiments with wine Pasteur had noticed how exposure to the air made some liquids ferment more quickly. Now he designed experiments to determine if something invisible in the air was truly causing the fermentation or if it were being caused by spontaneous generation within the liquid alone.

By placing bacteria-prone broths in special swan-necked vessels that could block out air, Pasteur was able to demonstrate that bacteria would grow only when air was allowed to pass through the neck of the vessel. He thus showed that it was invisible germs in the air, not spontaneous generation, that was causing the growth of bacteria on the liquids. As a further experiment, he exposed unfermented liquids to the pure mountain air of the high Alps. No new bacteria grew in this virtually germ-free environment.

Pasteur's early experiments with fermentation also suggested that heating a liquid to kill microorganisms could be applied to other perishable fluids. He discovered that by heating milk, for example, to a certain temperature for a certain amount of time, most of the bacteria present in it were killed. This process came to be known as pasteurization. Today dairies all over the world pasteurize foods, especially liquids. Pasteur himself developed pasteurization techniques for wine and beer production.

Discovers treatment for anthrax

During the 1870s Pasteur began work on human diseases. He first tackled anthrax, a potentially fatal animal disease of the skin that could be spread to humans. Building on discoveries made by German scientist Robert Koch (sometimes called the cofounder of microbiology), Pasteur succeeded in developing a vaccination that protected animals against anthrax. (A vaccine is a preparation that increases an organism's immunity, or antibody level, to a particular disease.) Like wine and silk, livestock were vitally important to the French economy. In a public test Pasteur infected sixty sheep, goats, and cows with anthrax. Previously he had vaccinated thirty-one animals; twenty-nine he left unprotected. Within a few days, all but four unprotected animals were dead, and the surviving four—all cows—were severely ill. None of the vaccinated animals showed the slightest sign of illness. Pasteur had won a dramatic victory.

Conducts important work with rabies

Pasteur's next discovery won him widespread acclaim. Thousands of people in the nineteenth century died every year from rabies, a fatal disease of the nervous system contracted from the bite of an infected animal. After experiments failed to show any bacteria at work in rabid animals, Pasteur realized the cause was another kind of germ (later called a virus) that was

Alice Evans, American Microbiologist

In 1918 American microbiologist Alice Evans (1881–1975) discovered the first confirmed case in which brucellosis, a disease common in cows, was transmitted to humans. The disease is very difficult to diagnose because its symptoms, which include fever, aches and pains, and inflammation, imitate those of other diseases such as influenza and typhoid fever. At first bacteriologists were skeptical of Evans's work. Some rejected her finding on irrelevant grounds—that she was a woman or that she had no Ph.D.—but others wondered how the transmission of brucellosis from cow to human had escaped scientists' notice for so long. It soon became obvious that chronic brucellosis was relatively common among families who drank raw milk of infected cows, but that it was typically diagnosed as another condition.

Evans herself developed chronic brucellosis as a result of her research, surviving a number of devastating attacks of the disease. She campaigned vigorously for the application of Pasteur's sterilization techniques in milk production. As a result of her work, vaccination of cows and pasteurization of milk became routine, bringing about dramatic declines in both cow and human diseases.

too small to be detected by a microscope. Following extensive work he succeeded in creating a vaccine against rabies.

The vaccine was effective in animals, but Pasteur was unwilling to risk experimenting on humans. Then in 1885 a boy named Joseph Meister, who had been bitten by a rabid dog, was brought to him. Since symptoms of rabies take several weeks to develop, Pasteur knew the vaccine would have time to take effect—and possibly save Meister's life—after the bite. Undoubtedly the boy would have died without it, so Pasteur injected him with the vaccine. He continued the injections over several days, and the child remained perfectly healthy. When Meister grew up, he became the gatekeeper at the Pasteur Institute, which Pasteur founded in Paris in 1888 to promote research on rabies. Pasteur remained the director of the institute until his death on September 28, 1895.

Further Reading

Dubos, Rene J., *Louis Pasteur: Free Lance Science,* Scribner, 1976.

Dubos, Rene J., *Pasteur and Modern Science,* Science Tech Publishers, 1988.

Latour, Bruno, *The Pasteurization of France,* HUP, 1988.

Newfield, Marcia, *The Life of Louis Pasteur,* Twenty-First Century Books, 1991.

Ruth Patrick

Born November 26, 1907
Topeka, Kansas

Ruth Patrick has pioneered techniques for studying the biodiversity (the numbers of different plants and animals) of freshwater ecosystems over a career that spans sixty years. Her studies of microscopic species of algae, called diatoms, in rivers around the world have provided methods for monitoring water pollution and understanding its effects. Federal programs that monitor the status of freshwater rely on Patrick's method of growing diatoms on glass slides. Her studies of the impact of trace elements (chemical elements present in minute quantities) and heavy metals on freshwater ecosystems (a community of organisms and its function as an ecological unit in nature) have demonstrated how to maintain a desired balance of different forms of algae.

"We're going to have to stop burning gasoline. And we're going to have to conserve more energy, develop ways to create electricity from the sun and plants, and make nuclear power both safe and acceptable."

Discovers importance of diatoms

Ruth Patrick was born in Topeka, Kansas, on November 26, 1907. She completed her undergraduate education at

IMPACT

In the late 1940s Ruth Patrick presented a paper at a scientific meeting on diatoms found in waters near the Pocono Mountains in eastern Pennsylvania. In the audience was an oil company executive, William B. Hart, who was so impressed with the possibilities of diatoms for monitoring pollution that he provided funds to support Patrick's research. Freed from financial constraints, Patrick undertook a comprehensive survey of the severely polluted Conestoga Creek, near Lancaster, Pennsylvania. It was the first study of its kind and launched Patrick's career. She matched types and numbers of diatoms in the water to the type and extent of pollution, an extremely efficient procedure now used universally.

Coker College, where she received a bachelor of science degree in 1929. She obtained both her master's degree, in 1931, and her Ph.D., in 1934, in botany from the University of Virginia in Charlottesville. The roots of Patrick's long and influential career in limnology (the study of freshwater biology) can be traced to the encouragement of her father, Frank Patrick. He gave his daughter a microscope when she was seven years old and told her, "Don't cook, don't sew; you can hire people to do that. Read and improve your mind."

Patrick's doctoral thesis, which she wrote at the University of Virginia, was on diatoms, whose utility derives from their preference for different water chemistries. The species of diatoms found in a particular body of water provide information about the makeup of the water. Yellow or brown in color, almost all diatoms are single-celled algae dwelling in fresh and salt water, especially in the cold waters of the north Pacific Ocean and surrounding Antarctica. Diatoms are an important food source for marine plankton (floating animal and plant life) and many other small animals. Diatoms have hard cell walls, called frustules, that are made from silica they extract from the water. (It is unclear how they accomplish this extraction process.) When the diatoms die, their glassy shells sink to the bottom of the sea and harden into rock called diatomite. One of the most famous and accessible diatomites is the Monterey Formation along the coast of central and southern California.

In 1933 Patrick joined the Academy of Natural Sciences in Philadelphia to work with one of the best collections of diatoms in the world. She had to volunteer her services because she was told at the time that women scientists were not paid. For income she taught at the Pennsylvania School of

John Cairns Jr., American Limnologist

American limnologist John Cairns Jr. (1923–), an associate of Ruth Patrick, is also an important figure in the environmental movement. Cairns has spent his entire career studying how ecosystems respond to stress, primarily to pesticides, industrial wastes, and other pollutants produced by human society. In 1948 Cairns began working at the Academy of Natural Sciences in Philadelphia as an assistant to Patrick. He eventually became assistant curator and then curator of limnology (the study of freshwater biology). In addition to studying the effects of the pollution stress on ecosystems—a field now known as "ectotoxicology"—Cairns has conducted extensive research on "restoration ecology," methods to heal stressed ecosystems. He has also examined ways in which freshwater microbial communities are structured and transported and how they develop and change. Using this information he was able to show in the early 1970s how to measure pollution effects in aquatic ecosystems in temperate (moderate temperature) zones. This is the chief method used for freshwater pollution studies in the People's Republic of China. In many of his writings Cairns cautions against uncontrolled population growth and other factors that continue to destroy natural systems. According to Cairns, a solution to the depletion of these systems can be found only if the members of all the scientific disciplines—wildlife biologists, agronomists (scientific farmers), and soil specialists, among others—combine their efforts.

Horticulture and made chick embryo slides at Temple University. In 1937 her persistence paid off, and she was appointed curator of the Leidy Microscopical Society with the Academy of Natural Sciences, a post she held until 1947. She also became associate curator of the academy's microscopy department in 1937. She remained in that capacity until 1947 as well, until she established the limnology department at the academy. She served as curator and chairperson of the department for more than four decades. Patrick headquartered her research in Philadelphia and had a field site in West Chester, Pennsylvania. An estuary (the mouth of a river where the tide of the sea flows in) field site at Benedict, Maryland, on the Patuxent River near Chesapeake Bay, serves as a base for studies of pollution caused by power plants.

Produces groundbreaking work

By her own account Patrick has waded into 850 different rivers around the globe in the course of her research. She participated in the American Philosophical Society limnological expedition to Mexico in 1947 and led the Catherwood Foundation's expedition to Peru and Brazil in 1955. As an advisor to several presidential administrations Patrick has given testimony on ecological problems and environmental legislation. She was an active participant in drafting the federal Clean Water Act. In 1987 Patrick addressed the growing problem of water contamination and depletion in the book *Groundwater Contamination in the United States.* Another of her concerns is global warming (the rise in Earth's temperature attributed to the buildup of carbon dioxide and other pollutants in the atmosphere), which many scientists believe is destroying the delicate balance of ecosystems worldwide.

Receives numerous awards

Patrick has received many awards, including the Gimbel Philadelphia Award for 1969, the Pennsylvania Award for Excellence in Science and Technology in 1970, the Eminent Ecologist Award of the Ecological Society of America in 1972, the prestigious Tyler Ecology Award in 1975, and the Governor's Medal for Excellence in Science and Technology in 1988. She holds many honorary degrees from U.S. colleges

Ruth Patrick in September 1983 using a Needham Scraper to dislodge sediment from a creek bed near Lafayette Hill, Pennsylvania.

and universities. Patrick has authored over 130 papers and, as a member of several governmental advisory committees, continues to influence thinking on limnology and ecosystems.

Further Reading

Detjen, Jim, "In Tiny Plants, She Discerns Nature's Warning on Pollution," *Philadelphia Inquirer,* February 19, 1989.

Patrick, Ruth, "Managing the Risks of Hazardous Waste," *Environment,* April 1991, pp. 13–35.

Patrick, Ruth, E. Ford, and J. Quarles, *Groundwater Contamination in the United States,* University of Pennsylvania Press, 1987.

Linus Pauling

Born February 28, 1901
Portland, Oregon
Died August 19, 1994
Big Sur, California

Over a seventy-year career, Linus Pauling earned two Nobel Prizes and left an enduring mark on the fields of science and politics.

Two-time Nobel laureate Linus Pauling—the only winner of unshared Nobel prizes in two separate categories—was the leading chemist of his era. His 1954 Nobel Prize for chemistry was given in recognition of his work on the nature of the chemical bond, and his 1962 Nobel Prize for peace was awarded for his efforts to bring about an end to the atmospheric testing of nuclear weapons. Rising from humble beginnings to international prominence, over the course of his seventy-year career Pauling came to have a profound influence on both science and society.

Linus Carl Pauling was born on February 28, 1901, in Portland, Oregon, the son of Herman Henry William Pauling and Lucy Isabelle (Darling) Pauling. Herman was a druggist who struggled to provide a decent living for his family, whom he moved through a series of small Oregon towns and eventually back to Portland in 1909. A year later he died of a perforated ulcer, leaving his wife to care for their three young children.

Pauling was a precocious child with a great appetite for books and a keen interest in science. Intrigued by a friend's chemistry set, he decided early on to become a chemical engineer. Pauling pursued his interest in science throughout his high school years, but, ironically, he didn't graduate with his class in 1917 because he failed to take the required courses in American history. Nonetheless, he entered the Oregon State Agricultural College (now Oregon State University) in Corvallis. Eager to pursue his study of science, Pauling registered for a full load of classes. Money was tight, however, and Pauling was forced to work full-time to support himself and to help his mother pay the household bills.

Studies with noted scientists

During his junior and senior years at Oregon State, Pauling learned about the work of physical chemist Gilbert Newton Lewis and chemist and physicist Irving Langmuir on the electronic structure of atoms and the way atoms combine with each other to form molecules. He became intrigued by the question of how the physical and chemical properties of substances are related to the structure of the atoms and molecules of which they are composed and decided to make this topic the focus of his own research. Also during his senior year Pauling met his future wife, Ava Helen Miller. The couple would marry in 1923 and would have four children.

Pauling received his bachelor's degree in chemical engineering in 1922 and then entered the California Institute of Technology (Cal Tech), a vigorous, growing institution that served as home to some of the leading researchers in chemistry and physics in the nation. After earning his doctorate in chemistry from Cal Tech in 1925, Pauling studied in Europe under noted physicists Arnold Sommerfield in Munich, Germany; Erwin Schrödinger in Zürich, Switzerland; and **Niels Bohr** (see entry) in Copenhagen, Denmark. These three scientists were all working in the new field of quantum mechanics, less than a decade old at the time. This science was based on the revolutionary concept that particles can sometimes have wavelike properties; these waves, in turn, were described as consisting of massless particles. Pauling was eager to apply this new way of looking at matter and energy to his own area of interest, the electronic structure of atoms and molecules.

After spending two years studying in Europe, Pauling returned to Cal Tech, where he accepted a position as assistant professor of theoretical chemistry. He wrote prolifically, publishing an average of ten papers a year during his first five years there. His reputation grew as well, and he was promoted quickly to full professor by 1930 and chairman of the chemistry department in 1936. Pauling would be affiliated with the institution for more than three and a half decades.

IMPACT

Linus Pauling's revolutionary research in molecular biology and biochemistry reshaped the entire scientific community's understanding of chemical bonding and helped establish a link between molecular irregularities and hereditary diseases. Pauling changed minds in the political arena as well; his support of a ban on nuclear testing was influential in the signing of a 1963 nuclear test ban treaty signed by the United States and the Soviet Union. For his diverse efforts Pauling was awarded two Nobel Prizes, one for chemistry in 1954 and one for peace in 1962.

Wins Nobel Prize for innovative work

In some ways the 1930s marked the pinnacle of Pauling's career as a chemist. As he devoted himself more and more to the study of the process of chemical bonding using the theories of quantum mechanics, he was able to apply quantum mechanical principles to solve a number of important problems in chemical theory. His first major paper on chemical bonding was published in April 1931, six more followed over the next two years, and in 1939 Pauling published his landmark book on the topic, *The Nature of the Chemical Bond and the Structure of Molecules and Crystals. The Nature of the Chemical Bond* has been considered one of the most important works in the history of chemistry, and the ideas presented in the book and the related papers are the primary basis upon which Pauling was awarded the Nobel Prize in chemistry in 1954.

Pauling's work on the chemical bond proved crucial in understanding three important problems in chemical theory: bond hybridization, bond character, and resonance. Hybridization refers to the process by which electrons undergo a change in character when they form bonds with other atoms. Carbon's electrons, for example, assume a new energy configuration during bonding.

In the bond character area, Pauling examined the relationship between two types of chemical bonding: ionic bonding, in which electrons are totally gained and lost by atoms, and covalent bonding, in which pairs of electrons are shared equally between atoms. Pauling was able to show that ionic and covalent bonding are only extreme states that exist in relatively few instances. More commonly, atoms bond by sharing electrons in some way that is intermediate between ionic and covalent bonding.

Pauling's work with resonance involved the problem of the benzene molecule. Until the 1930s the molecular structures that were written for benzene did not adequately correspond to the properties of the substance. German chemist Friedrich Kekulé had proposed a somewhat satisfactory model in 1865 by assuming that benzene could exist in two states and that it shifted back and forth between the two continuously. Pauling suggested, and quantum mechanics showed, that the most stable form of the benzene molecule was neither one of Kekulé's structures but some intermediate form. This intermediate form could be described as the superposition (the placement of one object over or above another) of the two Kekulé structures formed by the rapid interconversion (mutual conversion, or change between) of one to the other. This "rapid interconversion" was later given the name resonance.

Turns to biology and model building

By the mid-1930s Pauling's interest in molecular structure expanded to the field of biology. Biological molecules are complex substances that are found in living organisms and can contain thousands of atoms in each molecule rather than the relatively simple molecules that Pauling had studied previously, which contained only twenty or thirty atoms.

The first biological substance that Pauling turned his attention to was the hemoglobin molecule. (Hemoglobin is the substance that transports oxygen through the bloodstream.) In order to study hemoglobin, however, Pauling needed to know more about its structure, and the structure of proteins in general. He eventually developed a satisfactory molecular model for hemoglobin using a technique on which he frequently depended, model building. Pauling constructed atoms and groups of atoms out of pieces of paper and cardboard and then tried to fit them together in ways that would conform to quantum mechanical principles. Resembling something built of Tinkertoys, the model he determined for hemoglobin was that of a helix, a spiral-staircase-like structure in which a chain of atoms is wrapped around a central axis. Not surprisingly, Pauling's model-building technique was adopted by **James D. Watson** and **Francis Crick** (see entries) in their solution to the structure of DNA (deoxyribonucleic acid; the molecules housing essential genetic information in all living cells), a problem that Pauling himself very nearly solved.

Pauling also turned his attention to other problems of biological molecules. In 1939, for example, he developed the theory of complementarity (the quality or state of having the capacity for precise pairing) and applied it to the subject of enzyme reactions. (Enzymes are substances that bring about specific biochemical reactions in cells.) He later used the same theory to explain how genes might act as templates (patterns) for the formation of enzymes. In 1945 Pauling attacked and solved an important medical problem. By using chemical theory he demonstrated that the genetic disorder known as sickle-cell anemia is caused by the change of one single amino acid (the building blocks of proteins) in the hemoglobin molecule.

Fights for disarmament

The 1940s were a decade of significant change in Pauling's life. He had never been especially political and, in fact, had voted in only one presidential election prior to World War II. His wife, however, had long been active in a number of social and political causes. Perhaps because of her influence, and as a result of his own wartime research on explosions, Pauling became more concerned about the potential use of nuclear

Barry Commoner, American Biologist and Environmental Scientist

Known as "the granddaddy of environmentalists," American biologist Barry Commoner (1917–) worked with Linus Pauling and other scientists in the campaign to halt nuclear testing. Commoner is widely known as a writer and lecturer on the relationships among environmental, energy, and resource problems and their economic and political implications. Long before ecology became a national concern, Commoner was convinced that scientists had a moral obligation to inform the public about the potentially disastrous effects of some technological developments. He has led the scientific information movement in his longstanding efforts to alert the public to the effect of the "technosphere" on the biosphere.

In the 1950s Commoner joined Pauling in drafting a petition, eventually signed by eleven thousand scientists, that called on the United States and the Soviet Union to end atmospheric testing of atomic weapons. Commoner was also concerned about the fallout (the drifting of radioactive particles into the atmosphere) that resulted from nuclear explosions. Later he moved his focus to energy and broad environmental issues.

Besides working as an environmentalist, Commoner has had a long and distinguished career as a research scientist. His early work was on viral replication (how viruses reproduce), and he made basic discoveries regarding the roles of free radicals (molecules with unpaired electrons) in living things. An innovator, Commoner applied developments in various other fields to the study of biology. He founded and is the director of the Center for the Biology of Natural Systems (CBNS), now located at the Flushing campus of Queens College of the City University of New York. CBNS research topics include carcinogens (cancer-causing agents) in the environment, agricultural sources of pollution, organic farming, energy conservation systems for urban housing, solar energy, waste reduction and alternative methods of municipal waste disposal, and the origin of dioxin (a toxic substance) in incinerators.

weapons in future wars. He decided while on a 1947 boat trip to Europe that he would raise the issue of world peace in every speech he made in the future, no matter what topic.

From that point on, Pauling's speeches and published papers were devoted more to political issues than to scientific topics. A champion of worldwide nuclear disarmament, he offered scientific insights into the dangers of the rapid development of nuclear weapons and argued that continued weapons testing could put the biological welfare of future generations at risk. His 1958 book *No More War!* was an impassioned condemnation of the development and use of biological and chemical weapons, and it irked some of his colleagues, fellow citizens, and many legislators.

Later that year, armed with a petition bearing the signatures of more than eleven thousand scientists, Pauling vigorously urged the United Nations to ban nuclear weapons testing. He remained uncompromising in his criticism of international nuclear policy, and his strong opinions on disarmament only heightened the anticommunist fervor that existed in the United States in the 1950s. (The United States was in the midst of a cold war—a period of high military tension but no overt military action—with the communist Soviet Union.) He further clashed with the conservative establishment in the 1960s over his opposition to the Vietnam War, a military action in which the United States found itself entangled in an attempt to curb the spread of communism in southeast Asia. Neither professional nor popular disapproval could sway Pauling's commitment to the peace movement, however, and he and Ava continued to write, speak, circulate petitions, and organize conferences against the world's continuing militarism. In recognition of these efforts, Pauling was awarded the 1962 Nobel Peace Prize. The awards ceremony, held the following year, coincided with the signing of a limited nuclear test ban treaty between the United States and the Soviet Union.

Pauling pursued the cause of peace as a research professor at the Center for the Study of Democratic Institutions in Santa Barbara, California, from 1964 to 1967. (Pauling left Cal Tech in 1964.) He then moved on to the University of California at San Diego for two years, and afterward joined the staff of Stanford University. Following his retirement from teaching in 1974, he established the Palo Alto-based Linus Pauling Institute of Science and Medicine.

Prescribes vitamin C—in large doses

Pauling sparked considerable controversy in 1970 with the publication of *Vitamin C and the Common Cold,* a book that presents his views on the therapeutic value of ascorbic acid (vitamin C). The chemist believed that massive doses of vitamin C could boost the immune system and protect humans against a variety of infectious diseases. He also promoted the vitamin as an effective weapon in the control of cancers, certain viruses, and various mental disorders, although his claims have not yet been proven. Pauling took 18,000 milligrams of the vitamin each day, 300 times the recommended daily allowance. He died at his home near Big Sur, California, on August 19, 1994, at the age of ninety-three—two and a half years after being diagnosed with prostate cancer.

Further Reading

Chicago Tribune, August 28, 1994, sec. 2, p. 6.

Los Angeles Times, August 20, 1994, p. A1; August 21, 1994, p. A3.

New York Times, August 21, 1994, p. A1.

Omni, December 1986, p. 102.

Pauling, Linus, *No More War!* Dodd, Mead, 1958.

Pauling, Linus, *Vitamin C and the Common Cold,* W. H. Freeman, 1970.

People, September 5, 1994, p. 69.

Scientific American, March 1993, p. 36.

Serafini, Anthony, *Linus Pauling: A Man and His Science,* Paragon House, 1989.

Times (London), August 22, 1994, p. 17.

White, Florence Meiman, *Linus Pauling: Scientist and Crusader,* Walker & Co., 1980.

Ivan Pavlov

Born September 26, 1849
Ryazan, Russia
Died February 27, 1936
Leningrad, U.S.S.R.
(present-day St. Petersburg, Russia)

Russian physiologist Ivan Pavlov is best known for developing, through his experiments with dogs, the theory of conditioned reflex.

Russian physiologist Ivan Pavlov conducted research on mammalian digestion that earned him the Nobel Prize in 1904. Ironically, it was his work with the conditioned reflex that brought him lasting recognition. Indeed, the colloquial expression "Pavlov's dog" refers to Pavlov's famous experiments in which he taught a dog to salivate at the sound of a bell by associating the bell with feeding. This research led to a physiologically oriented school of psychology that focuses on the influence of conditioned reflexes on learning and behavior. Because of his contribution to the fields of psychology and physiology, Pavlov became one of Russia's most revered scientists and was even tolerated by the communist Soviet regime, of which he was openly critical.

Becomes interested in physiology

Ivan Petrovich Pavlov was born in Ryazan, Russia, on September 26, 1849, to Pyotr Dmitrievich Pavlov and Varvara

Ivanova. His father, who was a priest and a devoted scholar, taught his son at an early age to read all worthwhile books at least twice so he would understand them better—advice that helped shape Pavlov's intense dedication to his work. Since his parents also expected him to follow the family tradition of entering the clergy, he attended Ryazan Ecclesiastical High School and the Ryazan Ecclesiastical Seminary.

During his studies at the seminary Pavlov became interested in science, and physiology (the study of physical and chemical processes of an organism) in particular. He was greatly influenced by a radical philosopher named Dmitri Pisarev who supported many of the evolutionist theories of **Charles Darwin** (see entry). Darwin's theory of evolution states that gradual changes take place in a species over a long period of time. Through the process of natural selection, the species' strong traits are passed on to offspring while the weak traits disappear.

Leaves seminary to study natural sciences

In 1870, when the Russian government decreed that divinity students could attend nonsectarian (nonreligious) universities, Pavlov decided to leave the seminary and attend St. Petersburg University to study the natural sciences. At St. Petersburg, Élie de Zion, a professor of physiology, made a profound impression on Pavlov. After graduating in 1875 Pavlov followed Zion to the Military Medical Academy in St. Petersburg, where Zion had been appointed chair of physiology. Pavlov became an assistant in Zion's laboratory while working toward his medical degree. Pavlov's job didn't last long, since Zion was soon dismissed because he was Jewish. (At the time intense anti-Semitism, or anti-Jewish sentiment, was common in Russia and other parts of Europe.)

Protesting his mentor's dismissal, Pavlov left the Medical Academy and went to the Veterinary Institute, where he spent the next two years studying digestion and circulation. In 1877 Pavlov traveled to Breslau, Prussia (present-day Wroclaw, Poland), to study with Rudolf Heidenhain, a specialist in digestion. After receiving his medical degree in 1879, Pavlov went on to earn his postdoctoral degree and was awarded a Gold Medal for his dissertation in 1883.

Directs experimental physiological laboratory

After studying abroad on a government scholarship, Pavlov returned to Germany to work with Carl Ludwig on cardiovascular (relating to the heart and blood vessels) physiology and blood circulation. He also collaborated with Heidenhain on further digestion research. Another mentor, Sergei Botkin, eventually asked Pavlov to direct an experimental physiological laboratory. Devoted to "scientific medicine," the lab focused on physiological and pathological (abnormal) relations in an organism. It was at this time that Pavlov first developed his interest in "nervism," the pathological influence of the central nervous system on reflexes.

IMPACT

Ivan Pavlov's experimenting profoundly influenced biological research in a number of ways. He advanced a number of laboratory techniques, including the use of antiseptic (sterile) surgical procedures on laboratory animals and the conducting of long-lasting experiments on only one animal. He believed in minimizing an animal's pain for both moral and scientific reasons and led the way for the humane treatment of laboratory animals. His conditioned reflex theory—that a subject can be taught to automatically respond to a stimulus—became popular in the fields of psychiatry, psychology, and education. While some of Pavlov's early research has not stood the test of time and further scientific inquiry, he remains one of the fathers of modern science.

Receives Nobel Prize for digestion research

Pavlov's work on blood circulation had earned him a professorship at the Military Medical Academy. In 1895 he accepted the position of chairman of the physiology department at the St. Petersburg Institute for Experimental Medicine, where he spent the greater part of his remaining career and conducted his exhaustive research on digestion, which eventually gained him worldwide recognition in scientific circles.

Focusing on the physiology of digestion and of gland secretions, Pavlov devised an ingenious experiment in which he severed a dog's gullet (throat), forcing the food to drop out before it reached the animal's stomach. Through this simulated feeding, he was able to show that the sight, smell, and swallowing of food was enough to cause secretion of the digestive acidic "juices." He demonstrated that stimulation of the vagus nerve (one of the major nerves of the brain) influences secretions of the gastric glands (glands in the walls of the stomach that secrete gastric juice to aid in the digestion of food). Pavlov received the 1904 Nobel Prize in medicine or physiology for these pioneering studies on how the central nervous system effects digestion.

Earns professorship at the Military Medical Academy

Ironically, by the time Pavlov received the Nobel Prize his work was soon to be overshadowed. British physiologist William Bayliss and others demonstrated that chemical stimulation causes digestive secretions from the pancreas, the gland that secretes digestive enzymes and the hormones insulin and glucagon. (An enzyme is a substance that causes biochemical reactions in cells; a hormone is a chemical released in the body that produces a specific effect on targeted cells and organs.) Ever curious, Pavlov himself conducted experiments that also confirmed this discovery.

Conditions dog

Pavlov conducted his most famous studies after he received the Nobel Prize. Concentrating on digestion and the nervous system, he set out to determine whether he could turn normally "unconditioned" (naturally occurring) reflexes or responses of the central nervous system into "conditioned" (learned) reflexes. Pavlov noticed that laboratory dogs would sometimes salivate merely at the approach of lab assistants who often fed them. Through careful repeated experiments, Pavlov demonstrated that if a bell is rung each time a dog is given food, the dog eventually develops a conditioned reflex to salivate at the sound of the bell, even when food is not present. Thus, Pavlov showed that the unconditioned reflexes—gastric activity and salivation—could become conditioned responses triggered by a stimulus (the bell) not previously associated with

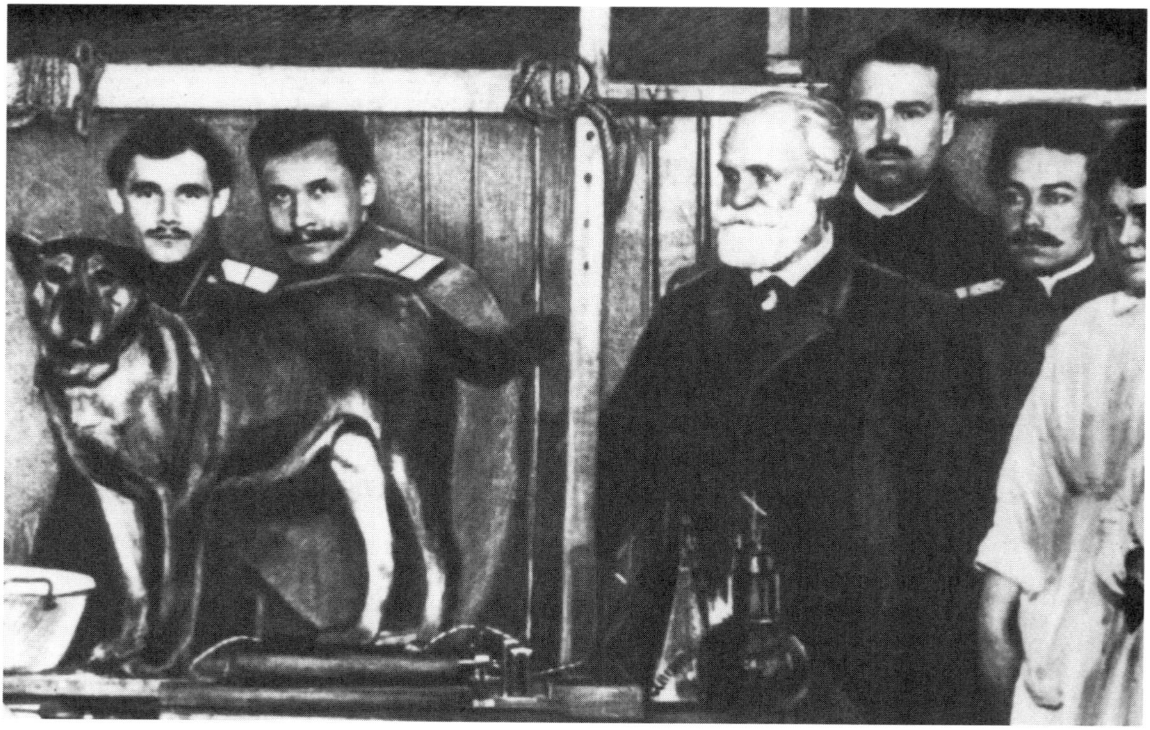

Ivan Pavlov and his staff demonstrating a conditioned reflex.

the physiological event (eating). Pavlov traced this phenomenon to the cerebral cortex and continued to study the brain's role in conditioned reflexes for the rest of his life. He published his masterwork on the subject, *Conditioned Reflexes,* in 1926.

Causes split in field of psychology

According to Pavlov, a person's behavioral development and learning are strongly affected by conditioned nervous responses to life events, similar to the dog's learned response to the bell. This idea, called behavioral theory, created a split in the field of psychology. Pavlov's followers opposed the views of **Sigmund Freud** (see entry), who theorized that an individual's thought processes—especially unconscious needs and desires—were the driving forces of human behavior. Freudian psychology, however, eventually surpassed Pavlovian psychology in popularity to become the primary approach to mental health treatment. But Pavlov maintained his devotion to the importance of conditioned reflexes in human behavior, believing that human language was probably the most intricate example of such conditioning. He also applied his theory to the treatment of psychiatric patients. He would place them in a quiet and isolated environment in order to remove any possible physiological or psychological stimuli that might negatively affect their mental health.

Protests communism

Pavlov's life spanned three distinct Russian political eras, which sometimes intruded upon his personal and professional life. He was born during the reign of Czar Nicholas I, a feudal monarch who sought to retain aristocratic (noble) rule at any price. Pavlov saw this oppressive regime give way to a new ideology of reform, known as post-Emancipation Russia, which heralded technological advancements but was mired in turmoil on both the social and political level. Shortly after the Bolshevik Revolution in 1917 which imposed a communist structure on the country—where the government forced the Russian people to give up their individual rights, including property—Pavlov became a staunch and vocal opponent of the new and often hostile regime.

By the time of the Bolshevik Revolution, Pavlov had achieved international fame and was living a comfortable life. (He had married in 1881 and had five children.) Yet he was willing to risk his hard-earned success by opposing the new communist regime. In 1924 he resigned from his position as chairman of the physiology department at the Medical Academy in protest of the expelling of all the clergymen's sons from the academy (the communists were opposed to organized religion). The new Soviet government was intent, however, on accommodating a person they considered to be a shining example of Russian science (even if that person were religious). As a result, Vladimir Lenin, the most powerful leader of the revolution, signed a decree guaranteeing Pavlov's personal freedom and his right to continue his research and even to attend church. These special privileges were in stark contrast to the harsh treatment given to countless other scientists whose work was suppressed by the government.

But the leaders of the new communist government had ulterior motives for keeping Pavlov safe. They believed that Pavlov's work with conditioned reflexes could be adopted for political purposes. They hypothesized that the public could be conditioned to accept communism and their rule just as Pavlov had conditioned the dog. In 1935, to appease their favorite scientist, the government built Pavlov a spacious laboratory equipped with the latest scientific technology. He called it "the capitol of conditioned reflexes."

Although Pavlov was known in the last five years of his life to publicly praise the government for their efforts to foster education and science, he repeatedly denounced the Soviet "social experiment." He died of pneumonia on February 27, 1936, in Leningrad.

Further Reading

The Great Scientists, Grolier Educational Corporation, 1989, pp. 186–91.

Vucinich, Alexander, *Science in Russian Culture,* Stanford University Press, 1963, p. 301.

Auguste Piccard

Born January 28, 1884
Basel, Switzerland

Died March 24, 1962
Lausanne, Switzerland

Jacques Piccard

Born July 28, 1922
Brussels, Belgium

Auguste Piccard was a scientist and inventor whose exploring instincts took him to record heights and depths. He and his twin brother, Jean, initially gained prominence together, but Auguste's accomplishments with his son Jacques were more sensational in nature and gained him a greater degree of popularity. Although Auguste and Jacques Piccard were neither meteorologists nor oceanographers, their technical achievements led to vital advances in both sciences.

Auguste Piccard reached new heights with his invention of a pressurized cabin for airships and new depths, with his son Jacques, in their bathyscaphe.

Auguste Piccard embarks on teaching career

Auguste Piccard was born into a prominent academic family in the Swiss city of Basel. He had a twin brother, Jean, who later became an important chemist. The brothers enrolled at the Swiss Federal Institute of Technology in Zürich; Auguste studied mechanical engineering and Jean chemical engineering. They both received their bachelor of science degrees and then went on to complete doctorates in their

> **IMPACT**
>
> For about 150 years, from the time Edmund Halley invented the first diving bell in 1716, oceanographers and engineers had been engaged in a series of efforts to overcome the obstacles of deep-sea exploration. The bathysphere, invented by William Beebe and Otis Barton in 1930, extended greatly a diver's depth range by protecting the diver from deep-water pressure. During the late 1940s and early 1950s Auguste and Jacques Piccard developed the bathyscaphe, which gave deep-sea explorers horizontal and angular mobility and independence from surface support vehicles. Today, sophisticated methods of remote sensing, such as those used by **Robert D. Ballard** (see entry) in his discovery of the *Titanic,* have enabled divers to explore the ocean more extensively and in more detail with unmanned expeditions.

respective fields. Between 1907 and 1920 Auguste taught in Zürich. He then accepted an appointment as professor of physics at the Brussels Polytechnic Institute, a post he held until his retirement in 1954. In 1919 he had married the daughter of a professor of history at the Sorbonne in Paris.

Designs special balloon

One of Piccard's earliest interests was Earth's upper atmosphere and the cosmic rays to be detected there. As early as 1783 scientists had been using lighter-than-air balloons to carry themselves and their instruments into the atmosphere to study its properties. Piccard and his brother made their first balloon ascension from Zürich in 1913, after which, in 1915, they both joined the balloon section of the Swiss army for a period of service.

The use of balloons to study the atmosphere was dangerous and limited; at a certain altitude the air becomes so thin that humans can no longer function. Around the turn of the twentieth century scientists launched dozens of unmanned balloons carrying instruments to record data more safely. Piccard maintained, however, that unmanned ascents could never provide the quality of data that could be obtained from balloons in which humans could travel. He therefore resolved to design a pressurized gondola in which observers could travel well beyond the 29,000-foot level that had marked the previous barrier to manned flight.

By 1930 Piccard's first design was ready for testing. The gondola was made of an air-tight aluminum shell that could be

pressurized to sea-level pressures and then suspended from a hydrogen-filled balloon. (The airtight cabin has since become a standard feature in airplanes.) On May 27, 1931, Piccard and a colleague, Paul Kipfer, took off in their airship from Augsburg, Germany. They eventually reached an altitude of 51,775 feet (9.8 miles), by far the highest altitude so far attained by human researchers. About a year and a half later, on August 18, 1932, Piccard made another record-breaking ascent, this time to a height of 53,139 feet (10 miles).

Auguste Piccard leans out of the air-tight gondola before his second balloon ascent on August 18, 1932. Piccard departed from Zürich, Switzerland, and rose more than ten miles into the air before landing near Venice, Italy.

Piccard made more than two dozen more balloon ascensions before he retired from the activity in 1937. During that time he collected valuable new information on atmospheric electricity and radioactivity as well as cosmic radiation. Probably more important, he continued to better the design of his aircraft, making the kinds of improvements that would eventually allow other scientists to reach altitudes of more than 100,000 feet (19 miles).

Invents bathyscaphe

In the late 1930s Piccard shifted his attention to the goal of reaching the lowest depth. He was convinced that the same techniques used to study the thin upper atmosphere could be used in the high-pressure depths of the sea. Using the same basic idea as his stratospheric balloon, he built a free-floating "deep-sea ship" that he called a bathyscaphe. The bathyscaphe consisted of two parts, an air-tight, pressurized cabin that was attached to a float that contained heptane (lighter-than-water gasoline) to provide the vessel with buoyancy. The bathyscaphe operated under its own power and could rise or sink by having seawater pumped into the flotation chamber or iron pellets dumped from the same chamber.

The first test of the bathyscaphe, which Piccard had named the *FNRS 2* for the Belgian scientific foundation that had supported the project, took place in 1948. The unmanned *FNRS 2* descended 4,600 feet (less than one mile), far less than Piccard had hoped. Another problem with this first bathyscaphe was that it could not be towed and had to be carried in a ship's hull. When the *FNRS 2* was damaged in heavy seas, Piccard set about building a better vessel. In the early 1950s he and his son Jacques, who had left teaching to work with his father full-time, designed and constructed the *FNRS 3*.

Jacques Piccard was born in 1922 and graduated from the École Nouvelle de Suisse Romande in Lausanne, Switzerland, in 1943. His graduate studies in economics at the University of Geneva were interrupted by World War II, in which he served in the French army from 1944 to 1945. He received his degree in 1946, taught for two years at the university, and then entered private teaching. Piccard soon left teaching to work with his father.

Piccards set depth records

In 1953, in their bathyscaphe *Trieste,* the Piccards traveled to a depth of 10,335 feet (almost two miles) off the coast of Capri in the Mediterranean Sea, a depth three times as great

as the previous record set in 1934. In 1954 Auguste retired from teaching and moved to Lausanne, Switzerland. (He died there on March 24, 1962.) In 1958 the U.S. Navy bought the *Trieste* and retained Jacques as a consultant. On January 23, 1960, Piccard and naval lieutenant Donald Walsh piloted the *Trieste* in a dive that went to a record depth of 35,802 feet—6.8 miles, a mile greater than the height of Mt. Everest—in the Mariana Trench in the Pacific Ocean. Piccard wrote about his extraordinary experience in the August 1960 issue of *National Geographic:* "Like a free balloon on a windless day, indifferent to the almost 200,000 tons of water pressing on the cabin from all sides, balanced to within an ounce or so on its wire guide ropes, slowly, surely, in the name of science and humanity, the *Trieste* took possession of the abyss, the last extreme on our earth that remained to be conquered."

The bathyscaphe Trieste *is lowered into Castellamare Harbor in Italy in preparation for an attempt at another world depth record in August 1953. Auguste Piccard made the descent inside the steel "bubble" protruding below the hull.*

Design mesoscaphe

Following the success of the bathyscaphe, the Piccards had developed the concept of the mesoscaphe, a "middle-depth ship" that would operate at depths down to 20,000 feet (3.8 miles). If the bathyscaphe could be compared to an underwater balloon, then the mesoscaphe was designed to be an underwater helicopter. In 1969 the Woods Hole Oceanographic Institute at Cape Cod, Massachusetts, carried out the Gulf Stream Mission using Jacques Piccard's mesoscaphe to conduct a series of experiments under the Atlantic Ocean.

During the mission Piccard made a month-long journey, from July 14 to August 14, 1969, traveling with the Gulf Stream from West Palm Beach, Florida, to a point 360 miles southeast of Nova Scotia. The six-man crew on the mesoscaphe measured and recorded the physical characteristics of the Gulf Stream and made observations of the rich animal life around them. The project also provided information on how men would react to confinement in a small space over a fairly long period of time, data that proved invaluable to the National Aeronautics and Space Administration's (NASA) development of its space program.

Further Reading

Cross, Wilbur, *Challengers of the Deep,* New York, 1965.

National Geographic, August 1960.

Field, Adelaide, *Auguste Piccard, Captain of Space, Admiral of the Abyss,* Houghton Mifflin, 1969.

Honour, Alan, *Ten Miles High, Two Miles Deep: The Adventures of the Piccards,* Whitlesey House, 1957.

Piccard, Auguste, *Entre terre et ciel,* Editions d'Ouchy, 1946, translation by Claude Apcher published as *Between Earth and Sky,* Falcon Press, 1950.

Piccard, Jacques, with Robert S. Dietz, *Seven Miles Down: The Story of the Bathyscaphe "Trieste,"* Putnam, 1950.

Max Planck

Born April 23, 1858
Kiel, Germany
Died October 4, 1947
Göttingen, Germany

German physicist Max Planck is best known as one of the founders of the quantum theory of physics, the theory that radiant energy is transmitted in the form of discrete units. After conducting research on heat radiation, he found that energy can sometimes be described in terms of separate units, which were later called quanta. Planck's discovery, now known as Planck's constant, was important because, for the first time, it was possible to use concepts related to matter (solid, liquid, or gas) to analyze energy. He also made important contributions in the fields of thermodynamics (the relationship between heat and other forms of energy), relativity, and the philosophy of science. For his quantum theory, Planck was awarded the 1918 Nobel Prize in physics.

"Today I have made a discovery which is as important as [Isaac] Newton's discovery."

Considers musical career

Max Karl Ernst Ludwig Planck was born on April 23, 1858, in Kiel, Germany. He was the youngest of Julius Wil-

IMPACT

German physicist Max Planck formulated a mathematical expression that explains the heat energy radiated by a black body. He thought of the energy as a stream of "energy bundles" called quanta, whose size are determined by the wavelength of radiation. His mathematical formula is simple: $E=h\nu$. E is the energy of the quantum; ν is the wavelength of the radiation; and h is a constant of proportionality (correspondence in size), which is now known as Planck's constant. The numerical value of Planck's constant can be expressed as 6.63×10^{-34} erg second (erg is a unit of work). The expression is engraved on the headstone of his grave at the Stadtfriedhof Cemetery in Göttingen, Germany. Today, Planck's constant is considered to be a fundamental constant of nature, much like the speed of light and the gravitational constant. Because of his discovery, the science of physics can now be subdivided into two great eras—classical physics, involving concepts worked out before Planck's quantum theory, and modern physics, consisting of ideas developed since 1900.

helm von Planck and Emma Patzig Planck's four children; his father also had two children from a previous marriage. Planck started school in Kiel, but when he was nine his family moved to Munich, where he attended the Königliche Maxmillian Gymnasium and graduated in 1874. At an early age Planck showed talent in a variety of areas, ranging from mathematics and science to music. An accomplished pianist and organist, Planck briefly considered a musical career, but he apparently abandoned that idea when a professional musician told him he lacked the necessary commitment. Planck maintained a life-long interest in music though, especially the mathematical foundations of musical composition. Later in his career Planck held private concerts in his home, performing on the piano with eminent musicians as well as fellow scientists, including the famous physicist **Albert Einstein** (see entry).

Studies physics in college

Planck entered the University of Munich in 1874, planning to major in mathematics. He soon changed his mind, however, when he realized he was intrigued by practical problems of the natural world instead of the abstract concepts of mathematics. Although he took courses that emphasized the practical and experimental aspects of physics, Planck was eventually drawn to theoretical problems. In 1875 he became ill and had to withdraw from the university. After a two-year recovery he transferred to the University of Berlin, where he studied with such notable physicists as Hermann von Helmholtz and Gustav Kirchhoff. By 1878 he was healthy enough to resume his studies at Munich. That year, after passing the state examination for higher level teaching in math and physics, he taught briefly at his alma mater, the Maxmillian Gymnasium. After teaching, Planck devoted his efforts to preparing his doctoral dissertation on the second law of thermodynamics (this law states that heat cannot move from a colder substance to a hotter one), earning a Ph.D. in 1879.

Marries and starts career

Planck's early research in thermodynamics was influenced primarily by the work of German physicist Rudolf Clausius. Planck had studied Clausius's work at Berlin and had analyzed it in his dissertation. From 1880 to 1892 Planck carried out a study of the principles of thermodynamics, especially their application to osmotic (absorbing or diffusing) pressure, boiling and freezing points of solutions, and the dissociation (chemical breakup into simpler units) of gases. The papers he published during this period were collected and published in his first major book, *Vorlesungen über Thermodynamik* (1897).

At the beginning of his career Planck was a privatdozent (lecturer) at the University of Munich. In 1885 he was appointed professor at the University of Kiel. His annual salary of two thousand marks (German dollars) allowed him to live comfortably and to marry Marie Merck, his childhood sweetheart from Munich. They would eventually have three children. Planck's personal life, however, was beset by tragedy when, in little more than three decades, he lost his entire family. His wife died in 1909, and by 1919 both of his twin daughters, Maragarete and Emma, had died while giving birth; his son Karl was killed in World War I (1914–18). After these devastating events, Planck married Marga von Hoessli, with whom he had a son, Erwin.

Investigates black body radiation

Planck's research on thermodynamics earned him recognition within the scientific community. Thus when Kirchhoff died

Wilhelm Wien, German Physicist

German physicist Wilhelm Wien's (1864–1928) investigations on black body radiation led Max Planck to formulate his quantum theory of physics. About 1893 Wien began a theoretical analysis of the characteristics of black body radiation, beginning with the fundamental laws of thermodynamics. He eventually developed two important conclusions. The first of these, now known as Wien's displacement law, states that the wavelength of radiation emitted by a black body is inversely proportional to the temperature of the body. That is, at low temperatures a black body will radiate energy with a long wavelength (red light). As the temperature rises, the most abundant wavelength radiated becomes smaller, and the color of the emitted light changes to orange, yellow, and then white.

Wien next attempted to find a mathematical formula for the relationship between the amount of energy radiated at each wavelength for various temperatures. He formulated a complex equation that works fairly well with short wavelengths, but not very well with long wavelengths. He published this result in June 1896. It was not until Planck introduced the concept of quantum energy in 1900, however, that the problem of black body radiation was finally solved.

in 1887, Planck was considered a worthy successor for his former teacher at the University of Berlin, and the following year Planck became an assistant professor at the university. In addition to his regular appointment he was also named director of the Institute of Theoretical Physics, which had been created especially for him. In 1892 Planck was promoted to the highest academic rank, ordinary professor, a post he held until 1926.

Originates quantum theory

At Berlin Planck turned his attention to the problem of black body radiation. A black body is any object that absorbs all frequencies (various levels of vibrating electric charges) of radiation (light) when heated and then gives off all frequencies as it cools. For more than a decade physicists had been trying to formulate a mathematical law that would describe the way in which a black body radiates heat.

The problem was unusually challenging because black bodies do not give off heat in the way scientists had predicted. Included among the explanations of this inconsistency were theories proposed by German physicist Wilhelm Wien (see box) and English physicist John Rayleigh. Wien's concept worked reasonably well for high frequency black body radiation, and Rayleigh's appeared to be satisfactory for low frequency. But no one theory adequately described black body radiation across the entire spectrum (range) of frequencies. Planck began working on the problem in 1896, and by 1900 he had found a solution. His solution depended on a groundbreaking assumption: the energy radiated by a black body is carried away in separate "packages," later given the name quanta (from the Latin *quantum,* for "how much").

Planck's concept was revolutionary because physicists had long believed that energy is always transmitted in some continuous form, such as a wave. They thought the wave, like a line in geometry, could be divided. Planck suggested that the heat energy radiated by a black body should be considered a stream of "energy bundles" whose size depended on the wavelength of the radiation. (A wavelength is the distance between one peak of a wave of light, heat, or energy and the next corresponding peak.) He formulated a simple mathematical formula for his concept: $E=h\nu$. E is the energy of the quantum; ν is the wavelength of the radiation; and h is a constant of proportionality (correspondence in size), which is now known as Planck's constant. Planck found that by making this assumption he could accurately describe the relationship between wavelength and energy radiated from a black body that had been observed in experiments. The problem was solved.

Wins Nobel Prize

Although Planck was a modest man, he recognized the significance of his discovery. According to one biographer, during a walk shortly after he had formulated the quantum concept Planck remarked to his son Erwin, "Today I have made a discovery which is as important as Newton's discovery." He was referring to the law of gravity formulated by English physicist Isaac Newton in the seventeenth century. In recognition of his achievement Planck was awarded the 1918 Nobel Prize in physics.

Ponders relativity and philosophical issues

After completing his study of black body radiation, Planck turned his attention to relativity, another new and important field of physics. In 1905 Einstein published a famous paper on the theory of general relativity that states that measurable properties will differ depending on the relative motion of the observer. This idea stimulated Planck to look for ways to apply quantum theory to Einstein's concepts. He was somewhat successful, especially in expanding Einstein's ideas from electromagnetism into mechanics. In another 1905 paper, this one discussing the production and transformation of light, Einstein himself made the first productive use of quantum theory in his solution of the photoelectric problem, also known as the photoemissive effect. The photoelectric effect involves the release of electrons from a metal that occurs when light (electromagnetic radiation) is shined on the metal. The problem was that the number of electrons released was determined not to be a function of the light's intensity, but of its color (that is, its wavelength). Einstein suggested that light be viewed as a collection of tiny bundles of energy rather than as a wave.

Throughout his life Planck had been interested in philosophical issues that extended beyond specific scientific research questions. As early as 1891 he had written about the importance of finding general themes in physics that could be used to integrate specific findings. Some of his books and papers were collected, translated, and published in 1959 as *The New Science.* Planck also contemplated ways science could be related to philosophy, religion, and society as a whole. He remained a Christian all of his life, frequently attempting to integrate his scientific and religious views. Like Einstein, Planck was never able to accept some of the fundamental ideas of modern physics that he had helped to create.

While he believed in God, he did not adhere to the idea of the deity having human qualities but instead found evidence of God in nature.

Experiences more tragedy in final days

By the time Planck retired from his position at Berlin in 1926, he had become probably the second most highly respected figure in Europe, if not the world (Einstein was the first). Four years after his retirement he was invited to become president of the Kaiser Wilhelm Society in Berlin, an institution that was then renamed the Max Planck Society in his honor. Because of his scientific status Planck was at first permitted to speak out against the increasingly repressive and anti-Jewish regime of Adolf Hitler, head of the National Democratic Socialist (Nazi) Party, which had taken control of Germany prior to World War II (1939–45). Nevertheless, his enemies managed to have him removed from his position at the Max Planck Society in 1937. Planck's final years were filled with additional personal tragedies. His son Erwin was found guilty of plotting against Hitler and executed in 1944. During an air raid by Allied forces (the United States, Britain, and Russia) on Berlin in 1945, Planck's home was destroyed along with all of his books and papers. Planck spent the last two and a half years of his life living with his grandniece in Göttingen, where he died on October 4, 1947.

Further Reading

Hermann, A., *Max Planck,* Hamburg, 1973.

Planck, Max, *The New Science,* Meridian Books, 1959.

Planck, Max, *Scientific Autobiography and Other Papers,* New York, 1968.

Weber, Robert L., *Pioneers of Science: Nobel Prize Winners in Physics,* American Institute of Physics, 1980, pp. 58–59.

Mark Plotkin

Born c. 1955
New Orleans, Louisiana

"Look, I'm just a city boy whose father sold shoes, and now Indians hug me in the jungle."

The world's tropical rain forests may hold a wealth of undocumented plant life that could be put to use as cures for disease. Finding and cataloging these helpful plants is the life's work of ethnobotanist Mark Plotkin, who is vice president for plant conservation at the Washington-based environmental group Conservation International. Plotkin has spent considerable time among native residents in remote areas of Suriname, Venezuela, French Guiana, and Brazil learning the medicinal properties of rain forest plants from community shamans (tribal priests) and medicine men. As Donald Dale Jackson put it in *Smithsonian* magazine, Plotkin and his associates "are hunting for pharmaceutical needles in a haystack that is shrinking steadily with the destruction of the rain forests and the erosion of tribal cultures." The scientist himself has helped to convince native rain forest peoples to preserve their jungle medicine traditions as both cultural treasures and potential profit-making ventures.

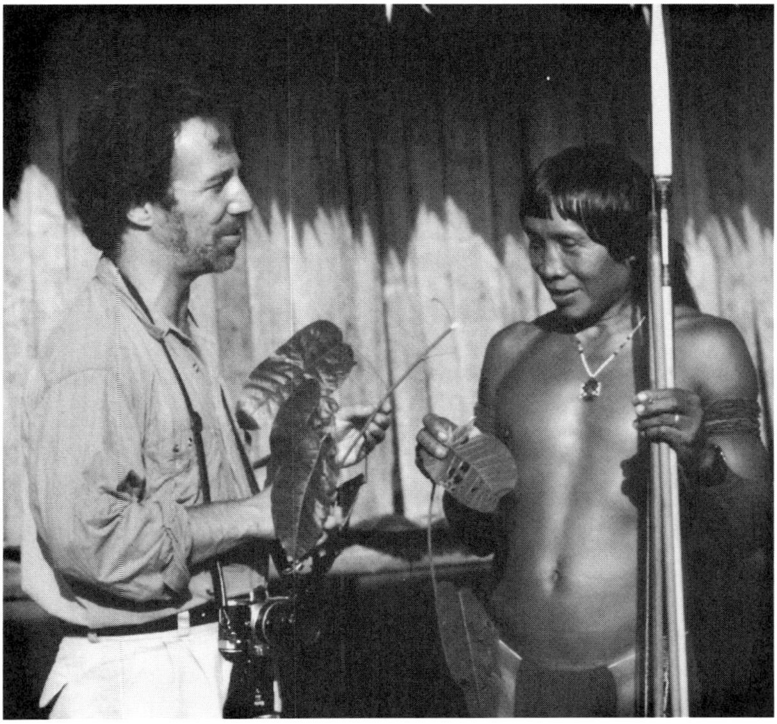

Mark Plotkin discusses medicinal uses of tropical plants with a Tirio Indian colleague in Suriname.

Always fascinated by nature

For Plotkin, the journey into an unusual career began in early childhood with a fascination for the natural world. According to a *Life* magazine article, Plotkin's first toys were plastic dinosaurs, and he learned the words "Tyrannosaurus Rex" shortly after he learned "Mommy" and "Daddy." In fact, "At age three, he asked his parents for a pet brontosaurus and was badly shaken to learn the meaning of the word 'extinct.'" In a sense Plotkin has been battling extinction ever since: both the loss of native customs brought about by the introduction of American goods and the destruction of the rain forest environment that might deprive humankind of valuable pharmaceuticals.

Collects rat snakes

Plotkin grew up in New Orleans, spending much of his spare time crawling through swamps collecting snakes and

IMPACT

As a vice president of the environmental group Conservation International, American ethnobotanist Mark Plotkin has devoted his life to fighting the destruction of the rain forests and to preserving the medicinal plants that it contains. He has undertaken dozens of excursions into remote jungle areas seeking information on native plant life and convincing the rain forest dwellers to preserve their native medicinal traditions. Plotkin wrote about his adventures in his 1993 book *Tales of a Shaman's Apprentice: An Ethnobotanist Searches for New Medicines in the Amazon Rain Forest.*

other wildlife. He told *Life* magazine that "there is no better feeling in the world than driving home after a long day, covered with mud, with a pillowcase full of rat snakes beside you." Armed with this fascination, Plotkin found college life at the University of Pennsylvania not to his liking—in the biology courses, he said, "everything was molecular and cellular" (dealing with life on the basis of molecules and cells). He dropped out of college and took a job working with the herpetology (a branch of zoology that deals with reptiles and amphibians) collection at the Museum of Comparative Zoology at Harvard University in Cambridge, Massachusetts. The position afforded him free tuition for night courses with the Harvard extension program, so he enrolled in a class called "Botany and Chemistry of Hallucinogenic Plants." That one course changed his life forever.

Decides on career

The course was taught by a renowned Harvard ethnobotanist (a scientist who studies plant lore and agricultural customs) named Richard Evans Schultes. On the first night of class Schultes showed a slide of three men wearing grass skirts and bark masks. He described them as Yukuna Indians doing a sacred dance under the influence of a hallucinogenic (mind-altering) potion—and then added dryly that the man on the left had a degree from Harvard. "That one slide did it," Plotkin recalled. "First of all, the rain forest in the background looked just like the pictures in my old dinosaur books. Second, it was wonderful to think of this strait-laced professor down in the jungle, wildly hallucinating on an Indian potion. Third, I wanted to save the world, and I realized that reptiles couldn't save the world, but *plants could.*"

Daniel H. Janzen, American Conservationist, Ecologist, and Biologist

American conservationist, ecologist, and biologist Daniel H. Janzen (1939–) may have the world's most ambitious gardening project: he is growing a tropical forest in Costa Rica. Since the mid-1960s Janzen has spent most of his time living in the 39-square-mile Santa Rosa National Park, located on Costa Rica's Pacific coast, just miles from the Nicaraguan border. With boundless zeal and curiosity, Janzen observes the minute detail of life among the 170 species of birds, 700 species of plants, 100 species of reptiles and amphibians, 115 species of mammals, and 13,000 species of insects, including more than 3,000 types of moths and butterflies, that he has identified at the park. His special interest is understanding how different species work together for ecological survival. In 1984 Janzen's work in what he calls "muddy-your-boot" biology won him the Crafoord Prize from the Swedish Royal Academy of Sciences, considered the Nobel Prize for natural sciences. In 1985 Janzen undertook another immense ecological experiment—regrowing the tropical dry forest that surrounds Santa Rosa.

Spends time in jungles

Plotkin earned a bachelor's degree from Harvard in 1979 and went on to obtain a master's degree from Yale University in New Haven, Connecticut, and a doctorate from Tufts University in Medford, Massachusetts. His Ph.D. dissertation concerned Surinamese ethnobotany among the Tirio Indian tribe. By the time Plotkin finished the study, he had made more than two dozen excursions into remote jungle areas seeking information on native plant life from elderly rain forest dwellers.

Plotkin told a *People* magazine correspondent that he has cataloged tribal shamans' use of "more than 300 plants to treat coughs, colds, skin disease, viruses, even diabetes." At one point the scientist was pleasantly surprised when a village doctor cured him of a bothersome fungal infection with tree sap.

At home in the United States, Plotkin has become an impassioned spokesman for preservation of rain forest environments and Native American cultures. Recently he outlined his adventures in the jungle and his hopes for igniting new interest in native medicines among a younger generation of Indians in the book *Tales of a Shaman's Apprentice: An Ethnobotanist Searches for New Medicines in the Amazon Rain Forest.* Plotkin also seeks new allies in the fight to conserve the rain forest from his position with Conservation International. "Rain forests contain well over half the world's plants," he told *Life.* "Who knows how many other medicines are out there that we haven't even discovered yet? I've spent my entire life loving jungles, and it just breaks my heart to see them and all their riches being lost."

When not on one of his field explorations or on a speaking tour, Plotkin lives in the Washington, D.C., area with his wife, who is also a conservationist, and their two daughters. His life, he told a *Smithsonian* magazine interviewer, is more than a dream come true. "Look, I'm just a city boy whose father sold shoes, and now Indians hug me in the jungle," he said. "These are forest Indians, and they don't *do* that. That's the dream of every kid who grows up wanting to be a jungle explorer."

Further Reading

Bender, Steve, "The Life Preservers," *Southern Living,* June 1995.

Life, June 1987.

New York Times Book Review, October 17, 1993.

Reed, Susan, "Sorcerers' Apprentice: What Amazon Shamans Know Could Save Your Life," *People,* December 6, 1993.

Plotkin, Mark, *Tales of a Shaman's Apprentice: An Ethnobotanist Searches for New Medicines in the Amazon Rain Forest,* Viking, 1993.

Jackson, Donald Dale, "Just Another Day in Paradise," *Reader's Digest,* April 1989.

Jackson, Donald Dale, "Searching for Medicinal Wealth in Amazonia," *Smithsonian,* February 1989.

Joseph Priestley

Born March 13, 1733
Fieldhead, England
Died February 6, 1804
Northumberland, Pennsylvania

English clergyman and chemist Joseph Priestly is credited with the discovery of oxygen and other gases.

English chemist Joseph Priestley made many important contributions to science, including the discovery of such gases as oxygen, nitrogen dioxide, and silicon fluoride. He also began experimenting with the gas known today as carbon dioxide. In addition to his work in science, Priestley was an activist whose writings on science, religion, and politics filled twenty-five volumes and whose support of the American and French Revolutions forced him to leave England.

Excels as a scholar

Joseph Priestley was born in 1733 into a poor family in Fieldhead, a village near Yorkshire, England. He was educated at religious schools that endorsed nonconformist beliefs, or doctrines that were opposed to those held by the Church of England. Although Priestly never formally studied science, he excelled as a scholar of languages, logic, and philosophy. At first Priestley became a country preacher, but eventually he turned to teaching. While employed at the Warrington Academy in the 1760s, he argued that school curriculums should reflect contemporary discoveries, rather than teach outdated classical models. In 1762 Priestley married Mary Wilkinson, the sister of one of his schoolmates.

Discovers carbon dioxide

In 1766 Priestley visited London, where he met the American statesman Benjamin Franklin, who was trying to settle a dispute between the American colonies and the British government. Soon after, Priestly moved to Leeds, where he took over a church parish and lived next door to a brewery. The layer of heavy gas that hovered over the brewery's huge fermentation vats aroused Priestley's scientific curiosity, and he began experimenting with this gas, known today as carbon dioxide. (In the fermentation process starches and sugars are broken down to form alcohol.) Finding that the gas was heavier than air and that it could extinguish flames, Priestley realized he had isolated the same gas that Scottish chemist Joseph Black had designated as fixed air (carbon dioxide). Conducting various experiments with this gas, he found that when dissolved in water, a bubbly drink was produced. He had invented soda water, or seltzer.

Identifies many gases

His interest piqued, Priestley devoted more time to chemistry experiments, particularly the study of gases. During the early 1770s he developed new methods of collecting gases in the laboratory and prepared several gases unknown to chemists at the time. He filled a common device called the pneumatic trough (a container for holding compressed air) with liquid mercury instead of water to obtain samples of gases such as sulfur, dioxide ammonia, and hydrogen chloride. He also worked with ammonia and hydrochloric acid, which were known earlier only as liquids, but which he isolated as gases. Priestley discovered nitrous oxide years before Sir **Humphry Davy** (see entry) popularized the properties of the gas. Other gases Priestly isolated and identified include nitrogen dioxide and silicon fluoride. As a result of these accomplishments, he was elected to the French Academy of Sciences.

Discovers oxygen

In 1773 Priestley won a lucrative post as librarian and companion to Sir William Petty, second Earl of Shelburne. Petty was a liberal politician who—like Priestly—sympathized with the plight of the American rebels. During his eight years with Petty, Priestley conducted his most famous scientific research. Because most of the gases he had studied were created by heating various substances, Priestley obtained a large magnifying lens which, in 1774, he used to discover oxygen. Although Swedish chemist Carl Wilhelm Scheele had discovered the gas a few years earlier, Priestley reported his results first and usually gets credit for the discovery. Priestley also found that mercuric oxide, when heated, breaks down to form shiny globules (tiny globes or balls) of elemental mercury while giving off a gas with unusual properties. A smoldering ember of wood, for example, burst into flames when exposed to the gas. Also, a mouse trapped in a container of the gas became frisky and survived for a longer time than it would have if trapped in ordinary air. And when Priestley inhaled the gas himself, he reported feeling "light and easy." He realized that this same gas was produced by plants, enabling them to restore "used-up" air to its original freshness.

In keeping with the scientific theory accepted at that time, Priestley named the gas "dephlogisticated air" because it absorbed phlogiston so readily. Phlogiston (the hypothetical material substance of fire) was thought to be the substance that gives materials their ability to burn or combust. Supposedly, during combustion, phlogiston is released from burning material and absorbed by the surrounding air or gas. When Priestley reported his findings, he unknowingly gave **Antoine Lavoisier** (see entry) the key to a new theory of combustion (a rapid chemical process that produces heat) that contradicted the phlogiston theory. Lavoisier later expanded on Priestley's work, renaming the gas oxygen and explaining how substances burn by combining chemically with oxygen.

Travels to America

Outside of science, Priestley was involved in radical religion and politics. During the 1770s and 1780s he wrote a number of controversial books, including *The History of the Corruption of Christianity,* which was published in 1782 and officially burned by the Church of England in 1785. His political views ran counter to what the British Crown found acceptable as well: he not only supported the American colonists' war with England, he sympathized with supporters of the French Revolution. (The French Revolution broke out in 1789 when peasants rose up against the oppressive French aristocracy; they eventually beheaded their king and queen and declared war on England.)

IMPACT

Joseph Priestley was responsible for several important advancements in the field of chemistry, some more well-known than others. In addition to his work with gases, Priestly was the first scientist to predict a relationship between electricity and chemistry, thereby anticipating the new field of electrochemistry. He also wrote two well-received science volumes, one on optics and *The History of Electricity,* published in 1767.

As political tension grew in England between supporters of the British monarchy and supporters of the French Revolution, Priestley's home and laboratory were burned down, destroying much of his research. Over time his religious allegiance had shifted toward the Unitarian Church (a form of Christianity that denies the doctrine of the Trinity, believing that God exists only as one person, not three), which was also unpopular in England. Priestley left for America in 1794. There he became a personal friend of Thomas Jefferson and other politicians. In fact, an essay on government published by Priestley in 1768 provided Jefferson with ideas for writing the Declaration of Independence. Priestley died in Northumberland, Pennsylvania, in 1804.

Further Reading

Heucher, James J., *Joseph Priestley and the Idea of Progress,* Garland, 1987.

Kieft, Lester, and Bennett R. Willeford Jr., *Joseph Priestley: Scientist, Theologian, and Metaphysician,* Bucknell University Press, 1979.

Edith H. Quimby

Born July 10, 1891
Rockford, Illinois
Died October 11, 1982

American biophysicist Edith H. Quimby made several important contributions in the field of radiology.

Edith H. Quimby was a leading American biophysicist, a specialist in the study of the mechanical principles of living tissues. As a pioneer in the field of radiology, she helped develop diagnostic and therapeutic applications for X rays, radium, and radioactive isotopes (isotopes are species of the same element with differing numbers of neutrons in their nuclei) when the science of radiology was still in its infancy. Her research in measuring the penetration of radiation, which causes the disintegration of atomic nuclei, enabled physicians to determine the exact dosages needed to treat diseases (usually cancerous tumors) with the fewest side effects for patients. Quimby also worked to protect those handling radioactive material from its harmful effects. While a radiology professor at Columbia University, she established a research laboratory to study the medical uses of radioactive isotopes, including their application in cancer diagnosis and treatment. In recognition of her contributions to the field, the Radiological Society of North America awarded her a gold medal for work that "placed every radiologist in her debt."

Begins career in scientific research

Quimby was born on July 10, 1891, in Rockford, Illinois, one of three children of Arthur S. Hinkley, an architect and farmer, and Harriet Hinkley. During Quimby's childhood the family moved to several different states. She graduated from high school in Boise, Idaho, then went on a full-tuition scholarship to Whitman College in Walla Walla, Washington, where she majored in physics and mathematics. Two of her teachers at Whitman, B. H. Brown and Walter Bratton, were major influences in directing her toward a career in scientific research.

After graduating in 1912 she taught high school science in Nyssa, Oregon, and then went to the University of California in 1914 to accept a fellowship in physics. While in the graduate program she married fellow physics student Shirley L. Quimby. She earned a master's degree in 1915 and returned to teaching high school science, this time in Antioch, California. In 1919, when her husband moved to New York to teach physics at Columbia University, Quimby went with him. The transfer to New York marked a pivotal point in Quimby's career, as she began working under Gioacchino Failla, chief physicist at the newly created New York City Memorial Hospital for Cancer and Allied Diseases. This began a scientific association that would last forty years.

Begins studying X rays

Quimby explored the medical uses of X rays and radium, especially in treating tumors. At that time physicians and researchers knew very little about this area of study. Before Quimby's research, for instance, individual doctors had to determine on a case-by-case basis how much radiation each of their cancer patients needed for treatment. Quimby focused her attention on measuring the penetration of radiation so that radiotherapy doses could be more exact and side effects could

IMPACT

Edith H. Quimby helped develop diagnostic and therapeutic applications for X rays, radium, and radioactive isotopes when little was known about the science of radiology. Prior to Quimby's discoveries doctors were quite inexperienced in radiation dosaging for cancer patients. Her findings on the effects of radiation on the skin and in the body led to significant advancements in radiotherapy treatments.

be minimized. After several years of research, she successfully determined the number of roentgens (a now obsolete unit of radiation dosage) per minute that were emitted in the air, on the skin, and in the body. Her investigation of the effects of radiation on the skin was especially noteworthy to the scientific community, and her study was frequently quoted in professional literature for many years.

Establishes isotope laboratory

From 1920 to 1940 Quimby conducted numerous experiments to examine various properties of radium and X rays. During this period she wrote dozens of articles for scientific journals, describing the results of her research and listing standards of measurement. In 1940 Quimby became the first woman to receive the Janeway Medal of the American Radium Society in recognition of her achievements in the field.

From 1941 to 1942 Quimby taught radiology courses at Cornell University Medical College. The following year she became associate professor of radiology at the Columbia University College of Physicians and Surgeons, where she taught radiologic physics. While at Columbia she and Failla, her mentor at New York City Memorial Hospital, founded the Radiological Research Laboratory, where they studied the medical uses of radioactive isotopes (different forms of the same element whose unstable nuclei break down and emit energy in the form of alpha, beta, or gamma rays) in cooperation with members of Columbia's medical departments. The pair focused its research on the application of radioactive isotopes in the treatment of thyroid disease (which adversely affects chemical reactions in the body), the study of circulation problems, and the diagnosis of brain tumors. These inquiries made Quimby a pioneer in the field of nuclear medicine.

Quimby participated in other aspects of radiology research as well. She explored the use of synthetically produced radioactive sodium (an element not occurring in nature but produced by means of nuclear reactions) in medical research and devoted considerable effort to providing protection to people handling dangerous radioactive substances. Quimby was among the first to foresee the potential for increased diagnostic and therapeutic use of atomic energy in medicine through radioactive isotopes.

Works on Manhattan Project

In addition to her research and lecturing, Quimby worked on the Manhattan Project, a secret, U.S. government-backed operation that developed the first atom bomb during World War II. She also worked for the Atomic Energy Commission, acting as a consultant on radiation therapy to the U.S. Veterans Administration; served as an examiner for the American Board of Radiology; and headed a scientific committee of the National Council on Radiation Protection and Measurements. A prolific writer, Quimby published a considerable amount of literature on the medical uses of X rays, radium, and radioactive isotopes. She also coauthored a highly respected book titled *Physical Foundations of Radiology.*

After her official retirement from Columbia University as professor emeritus of radiology in 1960, Quimby continued to write, lecture, and consult well into the 1970s. She was a member of several professional organizations, including the American Radium Society, for which she served as vice president. In her nonprofessional life, Quimby was a member of the League of Women Voters. She died on October 11, 1982, at the age of ninety-one.

Further Reading

New York Times, October 13, 1982, p. 28.

Physics Today, December 1982, pp. 71–72.

Charles F. Richter

Born April 26, 1900
Hamilton, Ohio
Died September 30, 1985
Pasadena, California

Charles F. Richter invented the Richter scale, the standard measure of earthquake intensity.

Charles F. Richter was an American seismologist (a scientist who studies earthquakes and the resulting shock waves that generate through the earth) best known for his 1935 invention of the Richter scale, which became the standard measure of earthquake intensity. Richter was a pioneer in seismological research at a time when data on the size and location of earthquakes was scarce. He authored two textbooks that are still used as references in the field of seismology and are generally regarded as his greatest contribution to science. Richter devoted his entire life to the study of seismology. At one time he even had a seismograph installed in his living room, and he welcomed queries about earthquakes at all hours of the day and night.

Applies physics to the study of the earth

Charles Francis Richter was born on April 26, 1900, on a farm near Hamilton, Ohio, north of Cincinnati. His parents were divorced when he was very young, and he grew up with his maternal grandfather who moved the family to Los Angeles in 1909. Richter went to a preparatory school associated with the University of Southern California, where he later spent his freshman year in college. He then transferred to Stanford University in California, earning a bachelor's degree in physics (the science that deals with energy and matter) in 1920. Eight years later Richter received his Ph.D. in theoretical physics from the California Institute of Technology (Cal Tech) in Pasadena. That same year he married Lillian Brand, a creative writing teacher. Taking a job at the newly established Seismological Laboratory in Pasadena, he started applying his physics background to the study of the earth. Richter dedicated the rest of his life to seismology.

Begins collaboration with Gutenberg

As a young research assistant Richter began a decades-long collaboration with Beno Gutenberg (see box), who was then director of the laboratory at Pasadena. In the early 1930s the two men were among several groups of scientists around the world who were trying to establish a standard way to measure and compare earthquakes. The seismological laboratory at Cal Tech was planning to issue regular reports on southern California earthquakes, so the Gutenberg-Richter study was especially important. They needed to be able to catalog several hundred quakes a year with an objective and reliable scale.

Develops the Richter scale

At the time the only way to rate shocks was with a scale developed in 1902 by Italian priest and geologist Giuseppe Mercalli. The Mercalli scale classified earthquakes from 1 to 12, depending on how people responded to the tremor and the extent of building damage. A shock that set chandeliers swinging might rate as a 1 or 2 on this scale, while one that destroyed huge buildings and created panic in a crowded city might count as a 10. The obvious problem with the Mercalli scale was that it relied on human perception and therefore was

IMPACT

Charles F. Richter, who is best known for his invention of the Richter scale, spent his career studying earthquakes. His theory that loss of life and property damage could be avoided by enforcing strict building codes—he opposed building anything higher than thirty stories in earthquake zones—was proven in Los Angeles, California, in 1971. The city suffered a major quake that year, and because city officials undertook preventive measures devised by Richter, many lives were saved. As a result of Richter's recommendations, there are no skyscrapers in Los Angeles.

not scientifically objective. It was also difficult to rate earthquakes that happened in remote, sparsely populated areas.

The scale developed by Richter and Gutenberg, which became known as the Richter scale, was an absolute measure of an earthquake's intensity. To record actual earth motion during an earthquake, Richter used a seismograph (an instrument consisting of a constantly unwinding roll of paper, anchored to a fixed place, and a pendulum or magnet suspended with a marking device above the roll). The scale takes into account the distance of the instrument from the epicenter, or the point on the ground that is directly above the earthquake's origin. Although the Richter scale immediately became the standard measure of earthquake intensity, it is still being used today in conjunction with the Mercalli scale in order to gauge the total impact of an earthquake. Since 1935 several other magnitude scales have been developed, including a recent one by Japanese-born American seismologist **Keiiti Aki** (see entry).

Locates epicenters of major earthquakes

For several decades Richter and Gutenberg worked together to monitor seismic activity around the world. In the late 1930s they applied their scale to deep earthquakes—those that originate more than 185 miles below the ground and rank particularly high on the Richter scale (8 or greater). In 1941 they published a textbook, *Seismicity of the Earth,* which in its revised edition became a standard reference book in the field. They worked on locating the epicenters of all the major earthquakes and classifying them into geographical groups. Richter warned, however, that seismological records reflect only what people have measured in populated areas and are not a true representation of all shocks that have actually occurred. He

Clarence Allen and Charles F. Richter assess seismographic readings of an earthquake that rocked California in February 1971.

long remained skeptical of some scientists' claims that they could predict earthquakes.

Saves many lives

Richter spent his entire career at Cal Tech, except for a visit to the University of Tokyo in Japan from 1959 to 1960 as

Beno Gutenberg, German-American Seismologist

German-born seismologist Beno Gutenberg (1889–1960) worked closely with Charles F. Richter to determine more accurate travel times of earthquakes and to clarify the relationships among magnitude, intensity, energy, and acceleration of vibrations in the earth. Gutenberg improved methods of determining the epicenter and depth of an earthquake, and his research with seismic waves provided invaluable insights into the internal layers of the earth. Investigations carried out in the 1930s on the travel time of seismic waves enabled him to locate the solid earth core, which produces waves more slowly than does the region containing molten matter that lies between the core and the outer crust. Gutenberg estimated the core to lie at a depth of 2,900 kilometers (almost 1,800 miles), a calculation that has not yet been disproved.

In 1947 Gutenberg was appointed director of the Seismological Laboratory at the Carnegie Institute. Continuing his own studies on the variations in the sizes of seismic waves, Gutenberg was able to make a more precise determination of the location of the layers through which waves travel relatively slowly. This channel, at a depth between 100 and 200 kilometers (about 60 to 120 miles), later turned out to play an important role in the study of plate tectonics (the movement of the earth's crust that is responsible for continental drift and causes earthquakes and volcanic activity). Gutenberg also maintained an interest in meteorology (the science that deals with weather), and his research on the structure of earth's upper atmosphere led to his ability to determine temperature patterns there.

a Fulbright scholar. He became involved in promoting earthquake building codes while at the same time discouraging the overestimation of the dangers of an earthquake in a populated area like Los Angeles. He pointed out that statistics reveal freeway driving to be much more dangerous than living in an earthquake zone.

Richter often lectured on how loss of life and property damage were largely avoidable during an earthquake with proper training and building codes—he opposed constructing any building higher than thirty stories in high-risk areas. In the early 1960s officials of the city of Los Angeles listened to Richter and began to remove potentially dangerous ornaments and cornices (horizontal, crownlike stone or brickwork) from its buildings. When the city suffered a major quake in February 1971 Richter was credited with saving many lives. He was also instrumental in establishing the Southern California Seismic Array, a network of instruments that has helped scientists track the origin and intensity of earthquakes as well as map their frequency much more accurately. His diligent study resulted in what has been called one of the most accurate and complete catalogs of earthquake activity, the Cal Tech catalog of California earthquakes.

Recalls actual earthquakes

Over the years Richter experienced several major earthquakes. He felt the 1933 Long Beach earthquake while working late one night at Cal Tech. That quake caused the death of 120 people in the then-sparsely populated southern California town; it cost the Depression-era equivalent of $150 million in damages. Nobel Prize-winning physicist **Albert Einstein** (see entry) was in town for a seminar when the Long Beach earthquake struck. He and a colleague of Richter's were crossing the campus at the time of the quake, so engrossed in discussion that they were apparently oblivious to the swaying trees. Richter also remembered the three great quakes that struck in 1906, when he was a six-year-old on the Ohio farm. That year San Francisco suffered an 8.3 quake on the Mercalli scale, while Colombia and Ecuador had an 8.9 and Chile had an 8.6.

Devotes entire life to seismology

In 1958 Richter published his text *Elementary Seismology,* which was derived from the lectures he faithfully taught to Cal Tech undergraduates and from decades of earthquake study. Many scientists consider this textbook to be Richter's greatest contribution to his field. In order to read international scientific papers in their original languages, he learned Russian, Italian, French, Spanish, German, and some Japanese. He even had a seismograph installed in his living room so that he could monitor quakes at any time.

During his lifetime Richter received considerable public and professional recognition. He was a member of the American Academy of Arts and Sciences and spent some time as president of the Seismological Society of America. After his retirement Richter helped start a seismic consulting firm that evaluated buildings for the government, public utilities, and private businesses. He pursued many interests, and is said to have been a devotee of the television series *Star Trek.* Richter died in Pasadena on September 30, 1985.

Further Reading

Los Angeles Times, October 1, 1985.

Pasadena Star-News, May 13, 1991.

Wilhelm Röntgen

Born March 27, 1845
Lennep, Germany
Died February 10, 1923
Weilheim, Germany

German physicist Wilhelm Röntgen discovered X rays.

During the first twenty years of his scientific career German physicist Wilhelm Röntgen studied a variety of topics, including the specific heats of gases, the magnetic effects of certain electrical materials, and the compressibility of water. He is most famous, however, for his discovery in 1895 of X rays, which had a revolutionary effect not only on physics but also on a number of other areas, particularly medicine. For this discovery Röntgen was awarded the first Nobel Prize in physics in 1901.

Expelled from school

Wilhelm Conrad Röntgen (also spelled Roentgen) was born in Lennep, Germany, on March 27, 1845, the only child of Friedrich Conrad Röntgen, a textile merchant, and Charlotte Frowein Röntgen. When Röntgen was three years old his family moved to Apeldorn in the Netherlands, fleeing a revolution in Germany. In Apeldorn Röntgen received his elementary and secondary education, and he also attended a private boarding school in the town of Middelann. In 1862 he enrolled at the Utrecht Technical School, but within two years his education was interrupted. Röntgen was unjustly accused of drawing a caricature (a picture with exaggerated features) of an unpopular teacher. Although another student was actually responsible for the drawing, Röntgen was expelled from school because of the incident. A few months later he was permitted to attend, on a trial basis, the University of Utrecht, where he took courses in scientific subjects such as physics, chemistry, and zoology. Röntgen's future still seemed bleak, however, because he had difficulty adjusting to the rigid Dutch educational system.

Finally one of Röntgen's friends told him about the more liberal admission policies at the Swiss Federal Institute of Technology in Zürich, Switzerland. Enrolling in the mechanical technical branch of the institute in 1865, Röntgen pursued a full course of study. However, distracted by the beauty of the surrounding mountains and lakes, he spent most of his time mountain climbing and boating. Only when one of his professors warned him that he would fail his examinations did Röntgen settle down and begin attending classes regularly. He received a diploma in mechanical engineering in 1868.

Begins graduate work

At Zürich Röntgen was influenced by the German physicist August Kundt, who suggested he do graduate work in physics (the science of matter and energy and the relations between them) rather than engineering. In 1869 he earned a doctoral degree for a thesis titled "Studies About Gases." Röntgen then became Kundt's assistant, and a year later he moved with Kundt to the University of Würzburg in Germany. While Röntgen was in Zürich he had met his future wife, Anna Bertha Ludwig, the daughter of a German revolutionary who had moved to Switzerland during the uprisings of 1848. They were married in 1872, after Röntgen moved to Würzburg. The couple would adopt Anna's six-year-old niece, Josephine Bertha, in 1887.

After two years at Würzburg, Kundt moved once more, this time to the newly established University of Strasbourg in France. Again Röntgen followed as his assistant. In 1874 Röntgen achieved a long-delayed ambition: he was appointed privatdozent (lecturer) at Strasbourg, his first official academic appointment, a position he had been unable to obtain in Germany because he lacked the necessary credentials. Over the next few years Röntgen took a series of academic appointments in France and Germany, eventually returning to Würzburg in 1888. In Würzburg he was named professor of physics and director of the university's Physical Institute. Röntgen would remain at Würzburg until 1900, serving as head of the university during his last six years.

During this period Röntgen wrote forty-eight scientific papers on diverse topics. His most significant contribution was a continuation of research originally suggested by **James Clerk Maxwell** (see entry) in his electromagnetic theory, which states through mathematical laws that electricity and magnetism cannot exist without one another. The theory also predicted that the motion of a dielectric (nonconducting) material within an electrostatic field (an area that consists of isolated, motionless charges of electricity) would induce a magnetic current within the dielectric material. In 1887, while working at the University of Geissen in Germany, Röntgen completed studies that confirmed this effect. As a result, Dutch physicist Hendrik Lorentz later named it "röntgen current."

IMPACT

Although Wilhelm Röntgen is credited with discovering X rays, he might not have been the first to observe them. Many earlier scientists, while experimenting with cathode-ray devices, which readily produce X rays, may have noticed but ignored such strange effects around their laboratories as glowing lights or overdeveloped photographic plates. Röntgen was the first scientist to recognize X rays as a new type of radiation, however, and the repercussions of his discovery spread rapidly. French physicist Henri Becquerel used X rays as the springboard of his own discovery of radioactivity; his work ultimately led to a better understanding of the atom, opening the door to the nuclear age. Scientists today consider Röntgen's discovery to be the beginning of the second scientific revolution, just as Italian physicist Galileo Galilei's establishing the scientific method—which stresses careful observation of a phenomenon and the formulation, testing, and confirmation of a hypothesis—started the first.

Experiments with cathode rays

Even though confirming Maxwell's theory was a significant scientific achievement, Röntgen is most famous for his discovery of X rays. In 1894 he began research on cathode rays, which was then one of the most popular topics in physics. Important work on cathode rays had been carried out in the 1870s by the English physicist William Crookes. He found that the discharge of an electrical current within a vacuum tube (an electron tube from which almost all the air has been pumped out) produces a beam of negatively charged rays. These rays cause a fluorescence (the giving off of light by a substance when it is exposed to electromagnetic radiation) on the glass walls of the tube. A number of scientists had followed up on Crookes's experiments, trying to find out more about the nature of his cathode rays.

In 1895, as Röntgen was repeating earlier experiments on cathode rays, his own research took an unexpected turn. In order to observe more clearly the luminescence (light generated by a process other than heating) caused by cathode rays, he darkened his laboratory and enclosed a vacuum tube in black paper. When he turned on the tube, he happened to notice that a nearby screen covered with barium platinocyanide crystals (a metallic substance) began to glow. This result was startling because Röntgen knew that cathode rays themselves travel no more than a few centimeters in the air. Therefore it had not been the rays that had caused the screen to glow.

Discovers X rays

Over a period of seven weeks Röntgen attempted to learn as much as possible about this new form of energy. Discovering that its effect could be detected at great distances from the vacuum tube, he concluded that the radiation was very strong. He found that the radiation passed easily through some materials, such as glass and wood, but was obstructed by other materials, like metals. At one point, as he held out a piece of lead between the tube and his hand, he even saw the bones in his

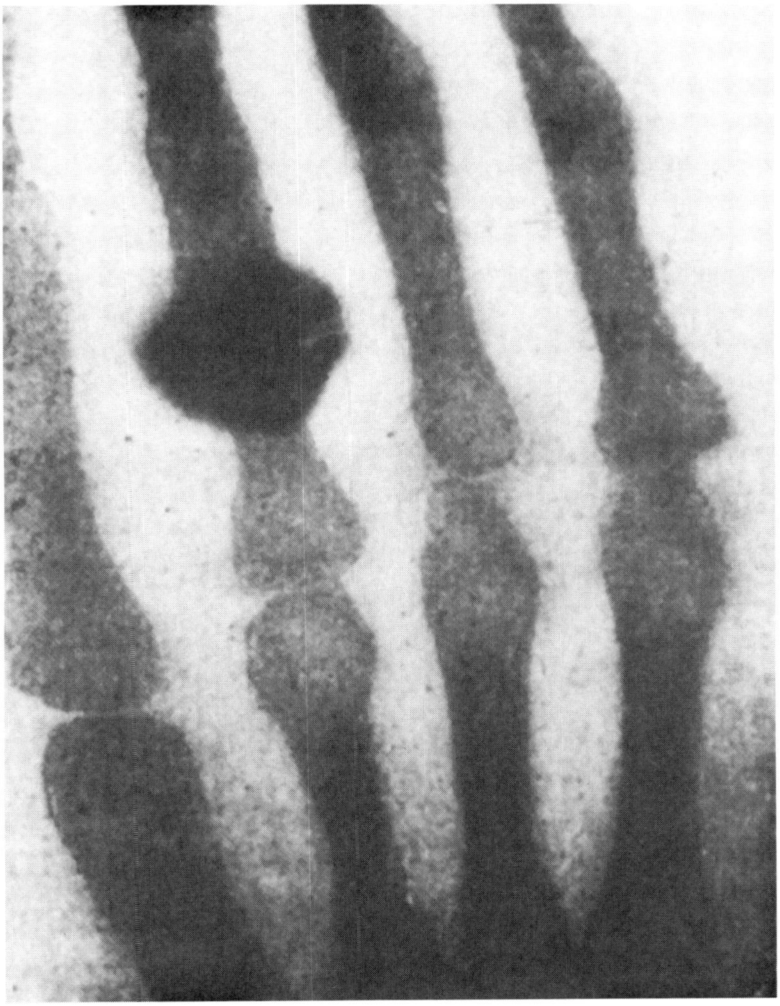

The first X-ray photograph, of the hand of Berth Röntgen, Wilhelm Röntgen's wife.

hand. He also discovered that the radiation was capable of exposing a photographic plate. Because of the unknown and somewhat mysterious character of this radiation, Röntgen gave it the name *X strahlen,* or X rays.

Captivates public with mysterious rays

In December 1895, soon after his first discovery of X rays, Röntgen sent news of his work to the editors of the scientific journal published by the Physical and Medical Society of

Würzburg. A few days earlier he had made the first X-ray photograph, a picture of his wife's hand. Fewer than twenty days later an X-ray machine was used in the United States to locate a bullet in a patient's leg. Within weeks news of Röntgen's discovery had reached the popular press, and the general public became captivated by the idea of seeing the skeletons of living people. Physicians proclaimed the discovery a modern miracle, while critics predicted an end to privacy, warning that X-ray devices could peer through walls, doors, and clothing. On January 13, 1896, Röntgen demonstrated his findings before the Prussian court (Prussia was a former kingdom of north-central Europe that included present day northern Germany and Poland) and was awarded the Prussian Order of the Crown, Second Class, by Kaiser Wilhelm II.

Wins first Nobel Prize in physics

After discovering X rays, Röntgen published two papers on the subject, one in 1896 and one in 1897, then returned to earlier experiments he had conducted on the effects of pressures on solids. He chose not to ask for a patent on his work, since he strongly believed that science belonged to everyone and that all nations should benefit from its advances. He also refused the Kaiser's offer of the honorary title "von" for his name, but he did accept the first Nobel Prize in physics in 1901. Even then, however, he declined to make an official speech and gave the prize money to the University of Würzburg for scientific research.

But Röntgen's discovery of X rays brought a number of personal attacks by many who said it was an accident or gave credit to other scientists. For several years Röntgen had declined offers from other universities, but in 1900, at the special request of the Bavarian government, he left his position at Würzburg to take a similar post at the University of Munich. The decision was not an easy one because he had a well-equipped laboratory at Würzburg. Röntgen remained at Munich until 1920 when he retired, partly as a result of his grief over his wife's death. She had suffered from a lingering illness and had become addicted to morphine, a drug extracted from opium and used in medicine as an anesthetic and sedative.

Becomes bankrupt in final years

The defeat of Germany in World War I (1914–18) had a devastating effect on Röntgen. The unstable economic period following the war resulted in his going bankrupt, and he spent his final years in poverty at his country home at Weilheim, near Munich. He died on February 10, 1923, from cancer, probably caused by prolonged exposure to X rays. (Prolonged exposure to X rays, a form of radiation, can cause cancer.) Among the awards presented to Röntgen during his lifetime were the Rumford Medal of the Royal Society (1896), the Bavarian Royal Order of Merit (1896), the Baumgartner Prize of the Vienna Academy (1896), the Elliott-Cresson Medal of the Franklin Institute (1897), the Barnard Medal of Columbia University (1900), and the Helmholtz Medal (1919).

Further Reading

Daintith, John, and others, *A Biographical Encyclopedia of Scientists,* Facts on File, Volume 20, 1981, p. 686.

Dictionary of Scientific Biography, Volume 1, Scribner, 1975, pp. 529–31.

Magill, Frank N., ed., *The Nobel Prize Winners—Physics,* Volume 1, *1901–37,* Salem Press, 1989, pp. 23–32.

Nitske, Robert W., *The Life of W. C. Röntgen, Discoverer of the X Ray,* University of Arizona Press, 1971.

Wasson, Tyler, ed., *Nobel Prize Winners,* H. W. Wilson, 1987, pp. 879–82.

Weber, Robert L., *Pioneers of Science: Nobel Prize Winners in Physics,* American Institute of Physics, 1980, pp. 7–9.

Bertrand Russell

Born May 18, 1872
Trelleck, Monmouthshire, Wales
Died February 2, 1970
Penrhyndeudraeth, Merionethshire, Wales

"Three passions, simple but overwhelmingly strong, have governed my life: the longing for love, the search for knowledge, and unbearable pity for the suffering of mankind."

An important figure in twentieth-century mathematical logic and philosophy, Bertrand Russell valued reason, clarity, and independence of judgment. He held the conviction that it was the duty of the educated and privileged classes to lead. Having protested against Britain's participation in World War I and opposed the development of nuclear weapons, Russell was imprisoned twice for his convictions. In writing his own obituary, he described himself as a man of unusual principles, who was always prepared to live up to them.

Influenced by grandmother

Bertrand Arthur William Russell was born on May 18, 1872, on the Russell family estate of Ravenscroft in the town of Trelleck in Monmouthshire, Wales. His father, John, Viscount Amberley, briefly served in the British Parliament but was defeated because of his rejection of Christianity, his advocacy of voting rights for women, and his support of birth control. Russell's mother was Katherine (Kate) Stanley, the daughter of a liberal politician. Orphaned before the age of five, Russell was raised by his paternal grandmother, a Scotch Presbyterian with strong moral standards, who gave duty and virtue greater priority than love and affection. Having no confidence in the moral and religious environment at boarding schools, she arranged for Russell to be taught by a series of governesses and tutors at home. Later he wrote in *The Autobiography of Bertrand Russell* that he had been influenced by his grandmother's contempt for convention and her indifference to the opinion of the majority. On Russell's twelfth birthday she gave him a Bible in which she had written her favorite texts, including "Thou shalt not follow a multitude to do evil." It was due to her influence, he felt, that in later life he was not afraid to champion unpopular causes.

Questions foundations of mathematics

When Russell was eleven his brother Frank began to teach him Euclidean geometry (a system of geometry based on axioms developed by Euclid, the ancient Greek geometer). "This was one of the great events of my life, as dazzling as first love," Russell recounted in his autobiography. "I had not imagined that there was anything so delicious in the world." Russell was disappointed to learn, however, that Euclid started with axioms that were not proved but were simply assumed to be valid. Russell refused to accept the axioms unless his brother could give a good reason for them. When his brother told him they could not continue their lessons unless he cooperated, Russell relented, but with reluctance and doubt about the foundations of mathematics. That doubt remained with him and determined the course of his future work in mathematics.

At age fifteen Russell left home to take a training course to prepare for the scholarship examination at Trinity College of Cambridge University. Lonely and miserable, he considered suicide but rejected the idea because, as he later said, he wanted to learn more mathematics. He kept a secret diary in which he questioned the religious ideas he had been taught. After a year

IMPACT

As Bertrand Russell was beginning his career, he formulated a plan to write a series of books on the philosophy of science. The first phase of the series would begin with mathematics and end with biology, and in the second phase he would address social and political questions. During his long life Russell managed to fulfill most of his plan, moving to the forefront of mathematical and social philosophy. According to his biographer Ronald Clark, no other twentieth-century Englishman equaled Russell in gaining respect from both the academic and nonacademic worlds.

and a half, he took the Trinity College examination and won a scholarship. One of the scholarship examiners was the English mathematician and philosopher Alfred North Whitehead (see box). Impressed with Russell's ability, Whitehead arranged for him to meet several students who soon became his close friends. Russell was also invited to join a group known as "The Apostles," which met weekly for intense discussions of philosophy and history.

After earning a first-class degree in mathematics in 1893, Russell found that his work in preparing for the exams had led him to think of mathematics merely as a set of tricks and ingenious devices, too much like a crossword puzzle. Disillusioned yet hoping to find a reason to believe in the truth of mathematics, he turned to philosophy. He studied idealism, a philosophy then popular at Cambridge, which contends that time, space, and matter are all illusions, and that the world resides in the mind of the beholder. Russell took the moral science examination in 1894, receiving an honors degree.

Studies German philosophers

At seventeen Russell had fallen in love with Alys Pearsall Smith, an American from a wealthy Philadelphia Quaker family. Although his grandmother did not approve, they became engaged after graduation. Hoping he would become interested in politics and lose interest in Smith, his grandmother arranged for him to become an attaché at the British Embassy in Paris. Russell was bored by his work at the embassy, however, and upon his return three months later he and Smith were married. In 1895 he wrote a dissertation on the foundations of geometry that won him a fellowship at Trinity College and enabled him to travel and study for six

Alfred North Whitehead, English-American Mathematician

Bertrand Russell collaborated with English-American mathematician Alfred North Whitehead (1861–1947), his former teacher at Trinity College, on *Principia Mathematica,* regarded as one of the most important books in mathematics ever written. Whitehead began his career publishing books on traditional mathematical topics, but eventually he shifted his attention to the philosophical foundations of science, writing several books on that topic. His main theme was that there exists a reality in the physical world distinct from the descriptions scientists have invented for that reality. Scientific explanations certainly have their functions, according to Whitehead, but they should not be construed as being the reality of nature itself.

After 1924, when he was appointed to the faculty at Harvard University, Whitehead extended the range of his writing to include even broader topics. His first work published in the United States, *Science and the Modern World* (1925), discussed the significance of science to other aspects of human culture. Whitehead gained an immediate reputation as a profound thinker and a writer of great clarity and persuasiveness.

years. He and his wife went to Berlin, Germany, where he studied the German socialist movement (a political and economic system through which government control of the production and distribution of goods would bring about a classless society) and the writings of Karl Marx.

Giuseppe Peano, Italian Mathematician

When Bertrand Russell began his analysis of the fundamental notions of mathematics, he used an approach developed by Italian mathematician and professor Giuseppe Peano (1858–1932). Peano is probably best known for the contributions he made to the development of symbolic logic. Indeed, many of the symbols he introduced in his research on logic are still used in science today. Some of Peano's most intriguing work involved the development of cases that ran counter to existing theorems, axioms, and concepts in mathematics. The most famous was his work on the space-filling curve (in which a graph of the curve covers every point within a given area), published in 1890. Peano's 1889 book *Arithmetices princepia, nova methoda exposita* ("The Principles of Arithmetic, Presented by a New Method") contains his postulates (assumptions) for the natural numbers (positive integers).

By 1900 Peano had become interested in an entirely different topic, the development of an international language. He saw the need for the creation of an "interlingua" through which people of all nations—especially scientists—would be able to communicate. Peano conceived of the new language as being the successor of the classical Latin in which pre-Renaissance scholars had corresponded, *a latino sine flexione,* or "Latin without grammar." He wrote a number of books on the subject.

Russell returned to London in 1896 and lectured at the London School of Economics and the Fabian Society (a socialist organization) on his experiences. That year he published his studies as *German Social Democracy,* which was the first of his seventy books and pamphlets. The Russells then traveled to the United States, where they visited Alys's friend, the poet Walt Whitman. Russell lectured on non-Euclidean geometry at Bryn Mawr College in Bryn Mawr, Pennsylvania, where his wife had been a student, and at Johns Hopkins University in Baltimore, Maryland. In 1900 he was invited to lecture on the works of German mathematician Gottfried Wilhelm Leibniz at Cambridge. Wanting to show that mathematical truths did not depend on the mathematician's point of view, he reexamined the philosophy of idealism. He ultimately abandoned idealism, concluding that matter, space, and time really did exist. Russell published his views in *Critical Exposition of the Philosophy of Leibniz.*

Writes The Principles of Mathematics

Russell next embarked on an ambitious task—devising a structure that would allow both the simplest laws of logic and the most complex mathematical theorems to be developed from a small number of basic ideas. If this could be done, then the axioms that mathematicians accepted would no longer be needed, and both logic and mathematics would be part of a single system. At a conference in Paris in 1900, Russell met Giuseppe Peano (see box), who had written a book in which he argued that mathematics was merely a highly developed form of logic. Russell became interested in the possibility of analyzing the fundamental notions of mathematics, such as order and cardinal numbers (numbers used in counting). Elaborating on Peano's approach, in 1903 Russell wrote *The Principles of Mathematics,* his thesis being that mathematics and logic are identical.

Formulates famous paradox

In *The Principles of Mathematics* Russell formulated his famous paradox (a self-contradictory statement that at first seems to be true), known as the Theory of Types. (He based his paradox on the proof developed by the Russian-born German mathematician Georg Cantor [see box], who stated that a class of all classes could not exist.) According to Russell, it is impossible to answer the question of whether the class containing all classes that are not members of themselves is a member of itself—for if it is a member of itself, then it does not meet the terms of the class. Likewise, if it is not a member of itself, then it does meet the terms of the class. (This is the same kind of paradox posed in the statement "Everything I say is a lie.") In developing the Theory of Types, however, Russell recognized the absurdity of asking whether a class can be a member of itself. He concluded that if classes belong to a particular type, and if they consist of homogenous (of the same or similar nature) members, then a class cannot be a member of itself.

Works with Whitehead

When planning *The Principles of Mathematics,* Russell had decided to write two volumes, the first providing explanations of his claims and the second containing mathematical proofs. Whitehead had been working on similar problems, so the two scholars decided to collaborate on the second part of the task. For nearly a decade they worked together, often sharing the same house, exchanging drafts of the book, and revising each other's work. The result, *Principia Mathematica,* became a separate work in three volumes, the last of which was published in 1913. Before the work was completed, Russell was elected to the Royal Society, a British scientific organization, of which Whitehead was already a member.

In 1907 Russell was persuaded to run for Parliament as a supporter of voting rights for women, but he lost the election. Three years later he was appointed a lecturer in logic and the principles of mathematics at Trinity College. In 1912 he published a short introduction to philosophy, *The Problems of Philosophy,* which he followed with *Our Knowledge of the External World* in 1914. That same year he presented the Lowell Lectures and taught a course at Harvard University in Cambridge, Massachusetts, and also lectured in New York City, Chicago, Illinois, and Ann Arbor, Michigan.

Goes to prison

An outspoken critic of England's participation in World War I, Russell worked with the No-Conscription Fellowship to protest the drafting of young men into the army. In 1916 he gave a series of lectures in London. The lectures were published as *Principles of Social Reconstruction* in England and appeared in the United States under the title *Why Men Fight: A Method of Abolishing the International Duel.* Russell was invited to give a lecture series at Harvard based on the book, but he was denied a passport because he had been convicted of writing a leaflet criticizing the imprisonment of a young conscientious objector (one who opposes war on moral grounds). As a result of his conviction he was dismissed from his lec-

tureship at Trinity College. Russell then wrote an open letter to Woodrow Wilson, the president of the United States, urging him to seek a negotiated peace rather than go to war. Since Russell's mail was being intercepted by the British government, he sent the letter with a young American woman. The story made the headlines in the *New York Times.*

Russell wrote for *The Tribunal,* the publication of the No-Conscription Fellowship. In one article he predicted that the consequences of the war in England would include widespread famine and the presence of American soldiers who would intimidate striking workers. Charged with making statements likely to prejudice Britain's relations with the United States, Russell was sentenced to six months in prison. Before his imprisonment he wrote *Roads to Freedom: Socialism,*

Bertrand Russell romping with children—including two of his own—at his experimental school in Hampshire, England, in March 1931. Despite such novel practices as allowing the children to attend only when they wished, never physically punishing the children, and allowing them to be rude and to go unclothed in hot weather, the school was a great success.

Anarchism and Syndicalism. While in prison he wrote *Introduction to Mathematical Philosophy,* which explained the ideas of *The Principles of Mathematics* and *Principia Mathematica* in relatively simple terms.

Travels extensively

In 1920 Russell visited Russia, where he became disappointed with the results of the 1917 communist revolution, a movement that began with high ideals but turned bloody and oppressive. He also lectured at the National University of Peking in China. The following year Russell divorced his wife and married Dora Black; they had two children, John Conrad and Katharine Jane. In 1927 the Russells established the experimental Beacon Hill School in Hampshire, England, which would influence the building of similar schools in Britain and the United States. During the late 1920s and early 1930s Russell traveled to the United States, speaking on political and social issues. In 1935 he divorced Dora and the following year married Patricia Spence. A son, Conrad Sebastian Robert, was born in 1937.

During World War II, from 1938 to 1944, Russell lived in the United States, lecturing at various universities throughout the country. He was invited to teach at New York City College, but the invitation was revoked because of objections to his atheism and unconventional personal morality (Russell had written a 1929 worked titled *Marriage and Morals* in which he expressed his views on divorce, adultery, and homosexuality, positions that were considered unconventional at the time). Russell continued to publish works on philosophy, logic, politics, economics, religion, morality, and education. In 1944 he returned to Trinity College, where he had been offered a fellowship. He published *History of Western Philosophy,* participated in radio broadcasts in England, and lectured in Norway and Australia. Russell was awarded the Order of Merit by King George III of England in 1949 and the Nobel Prize for literature in 1950. At the age of eighty he divorced his third wife and married Edith Finch.

Georg Cantor, Russian-born German Mathematician

Russian-born German mathematician Georg Cantor's (1845–1918) proof that a class of all classes cannot exist inspired Bertrand Russell to formulate his famous paradox concerning classes. Cantor developed a number of other ideas that profoundly influenced twentieth-century mathematics. His accomplishments earned him recognition as the founder and creator of set theory (the mathematical study of the properties of sets), and he made significant contributions to classical analysis. In addition, he did innovative work on real numbers (members of rational or irrational number sets) and was the first to define irrational numbers (numbers such as square roots) by sequences of rational numbers (whole numbers or fractions). Renowned mathematician **David Hilbert** (see entry) praised his work, claiming, "Cantor has created a paradise from which no one shall expel us." Nevertheless, Cantor's revolutionary concepts were accepted only gradually and not without opposition during his lifetime.

Campaigns for nuclear disarmament

Acutely aware of the dangers of nuclear war, Russell served as the first president of the Campaign for Nuclear Disarmament in 1958, and as president of the Committee of 100 in 1960. As a member of the Committee of 100 he encouraged demonstrations against the British government's nuclear arms policies. For this activity he was sentenced to two months in prison, although the term was reduced to one week because of his poor health. In his ninth decade Russell established the

Bertrand Russell Peace Foundation, published *The Autobiography of Bertrand Russell,* and appealed on behalf of political prisoners in several countries. He also protested nuclear weapons testing, criticized the war in Vietnam (which escalated with U.S. involvement in an attempt to curb the spread of communism in southeast Asia), and established the War Crimes Tribunal (an international court that tried people accused of military atrocities). Russell died on February 2, 1970, at the age of 98.

Further Reading

Clark, Ronald, *Bertrand Russell and His World,* Thames & Hudson, 1981.

Russell, Bertrand, *The Autobiography of Bertrand Russell,* three volumes, Little, Brown, 1967–69.

Ernest Rutherford

Born August 30, 1871
Spring Grove, New Zealand
Died October 19, 1937
Cambridge, England

New Zealand-born English physicist Ernest Rutherford won the 1908 Nobel Prize in chemistry for his explanation of radioactivity. His most renowned achievement, however, was his classic demonstration that the atom consists of a small, dense nucleus surrounded by orbiting electrons. Rutherford also showed how one element can be changed into another by splitting the atom. Under Rutherford's direction, laboratories in Canada and Great Britain discovered the neutron and helped launch high-energy physics, which is also known as particle physics.

New Zealand-born English physicist Ernest Rutherford discovered that the atom consists of a nucleus surrounded by orbiting electrons.

Ernest Rutherford was born on August 30, 1871, to James and Martha (Thompson) Rutherford near the village of Spring Grove, New Zealand. His early success in school earned him a scholarship to Nelson College, a secondary school in a village on the north end of New Zealand's South Island. He then received a scholarship to Canterbury College at Christchurch, New Zealand, where he earned his bachelor's degree in 1892. He continued studying at Canterbury, earning a master's

IMPACT

Ernest Rutherford's explanation of radioactivity at the atomic level excited the scientific community and won him the 1908 Nobel Prize in chemistry. While conducting his groundbreaking atomic structure research he also discovered that the atom consisted of a small dense nucleus surrounded by orbiting electrons. This discovery led to the production of the model of the atom used today.

degree with honors in mathematics and mathematical physics in 1893 and a bachelor of science degree in 1894. During this time Rutherford met Mary Newton, whom he married in 1900. The couple would have one daughter, Eileen.

Researches electromagnetism

During the course of his studies Rutherford researched the effects of electromagnetic waves on the magnetization of iron. (Electromagnetic waves are forms of energy that combine electricity and magnetism and are produced by rapidly alternating electrical currents; magnetism is the property of a body to produce an electrical current around itself.) He observed that, contrary to contemporary expectations, iron did magnetize in high-frequency electromagnetic fields. Conversely, he also showed that electromagnetic waves could demagnetize magnetized iron needles. On the basis of this observation, Rutherford devised an electromagnetic detector, a device for picking up electromagnetic waves produced at a distance. Italian physicist **Guglielmo Marconi** (see entry) would later use the same principles in the development of wireless telegraphy, or the radio.

In 1895 Rutherford earned a scholarship to Trinity College at Cambridge University in England. He worked as the university's first research student under the direction of renowned English physicist J. J. Thomson at the Cavendish Laboratory. The laboratory had been established in 1871 for research in experimental physics and was first led by electromagnetism pioneer **James Clerk Maxwell** (see entry). Rutherford's demonstration of the electromagnetic detector immediately impressed Thomson and other scientists at the Cavendish.

In 1896 Thomson invited Rutherford to assist him in studies of how X rays, which had been discovered in 1895 by **Wil-**

helm Röntgen (see entry), affect the electrical properties of gases. Thomson and Rutherford demonstrated that X rays break gas molecules into electrically and positively charged ions (an ion is an atom that has either gained or lost one or more electrons), making the gas electrically conductive (able to transmit an electrical current). This work brought Rutherford widespread recognition in the British scientific community.

Contributes to understanding of radioactivity

In 1897 Rutherford took up the study of radioactivity—the emission of radiation, or energy emitted in the form of waves—which had been discovered almost accidentally by French physicist Henri Becquerel the year before. Rutherford began by experimenting with the radioactive emissions of uranium, a silvery-white metallic element found in several minerals. In order to observe the penetrating ability of these emissions, he systematically wrapped uranium in layers of aluminum foil. He concluded that uranium emitted two distinct types of radiation: a less penetrating type, which he called "alpha radiation," and a more penetrating type, which he called "beta radiation." Later Rutherford also observed the most penetrating type of uranium, which was described by French physicist Paul Villard as "gamma radiation."

In 1898 Rutherford took the position of Second MacDonald Professor of Physics at McGill University in Montreal, Canada. At this time he began studying another radioactive element called thorium. Although thorium, like uranium, emits alpha and beta radiation, emission patterns for thorium substances seemed erratic (nonuniform). By the following year Rutherford had determined that the thorium was producing an emanation (a new radioactive substance). He also observed that the emanation's radioactivity gradually decreased geometrically with time, an occurrence now known as the half-life of a radioactive substance. (Half-life is a measurement of the time it takes for half of a substance to decay.)

Rutherford joined Frederick Soddy, an Oxford chemistry demonstrator, in 1901 to study the thorium emanations and to

Hans Geiger, German Experimental Physicist

After radioactivity was discovered in 1896, many scientists actively pursued the development of instruments to detect and count various types of radiation, which cannot be detected by the human senses unaided. German experimental physicist Hans Geiger (1882–1945) constructed the earliest form of the radiation counter that bears his name in 1908. The device consisted of a wire extending down the center of a sealed metal tube with a glass or mica (a mineral highly resistant to heat) window at one end. The wire and the metal tube were connected to a power source through an external circuit. When radiation passed through the tube, it created a track of ion pairs. The positively charged ions in each pair were attracted to the negatively charged metal tube, while the negatively charged electrons were attracted to the central wire. As the ions and electrons passed through the gas, they collided with other gas molecules, producing further ionization. Because of the avalanche effect, the number of ions and electrons reaching the outside circuit was sufficient to initiate an electric current that could be recorded as a sound or observed as a flash of light.

Geiger continued to work on the design of his counter over the next two decades. In 1928 he made further modifications based on the studies of a German colleague, Walther Müller. The final design, often referred to as the Geiger-Müller counter, soon became one of the most widely used radiation counters available to scientists.

explain curious observations of other radioactive substances. These substances had been noticed in Europe by Becquerel and by English physicist and chemist Sir William Crookes, who discovered thallium. Both had isolated (separated for the purpose of studying) the active parts of uranium from an apparently inert (inactive) part. However, Becquerel also observed that the active part soon lost its activity, while the inert remainder regained its activity.

Rutherford and Soddy isolated the active part of radioactive thorium, which they named thorium-X, from the apparently inert parent thorium. They charted how thorium-X gradually lost its radioactivity, while the original thorium regained its activity. Their experiments illustrated that thorium-X had its own distinctive half-life, which was much shorter than the half-life of thorium. From these observations Rutherford and Soddy established the modern concept of radioactivity in 1903.

Rutherford and Soddy's explanation of radioactivity at the atomic level caused a sensation in scientific circles. Rutherford showed that radioactivity—alpha, beta, and gamma radiation—was the physical manifestation of this disintegration, the pieces of the thorium atom that were released as it decayed. In other words, thorium was steadily being transmuted (transformed) into a new element that was lower in atomic number. (Atomic number is the number of protons in the nucleus of an atom.) For this work Rutherford would earn the 1908 Nobel Prize in chemistry.

Becomes leading authority on radioactivity

In 1903 Rutherford became a fellow of the Royal Society, a British scientific organization. The following year he received the prestigious Rumford Medal. His books on radioactivity became standard texts on the subject for years, and he was a popular speaker. He attracted a number of talented associates at McGill, including German physicist Otto Hahn who would, with Austrian physicist **Lise Meitner** (see entry), demonstrate the fissioning (splitting) of uranium in 1939.

Studies alpha particles

In 1907 Rutherford was appointed professor of physics at Manchester University in England. At Manchester he had a well-equipped laboratory and worked with talented associates from around the world, such as physicists **Niels Bohr** (see entry), Hans Geiger (see box), Ernest Marsden, and H. G. J. Moseley, as well as Charles Darwin, grandson of the famous naturalist. Rutherford continued with his study of radioactive emissions, particularly the high-energy alpha particles, because he wanted to determine the precise nature of the alpha particles. In 1903 at McGill Rutherford had succeeded in

Hans Geiger and Ernest Rutherford at the Schuster Laboratory.

deflecting (scattering) alpha particles in electric and magnetic fields, proving they had a positive charge. He was certain that the relatively massive particle must consist of two protons like a helium atom's nuclei. At Manchester in 1908 Rutherford and his colleagues experimentally proved that the alpha particles were indeed nuclei of helium atoms.

For this experiment Rutherford and Geiger devised a method for precisely counting alpha particles by firing them into a nearly evacuated tube that had a strong electric field. The resulting ionizing effect (conversion into ions) in the gas could be detected by an electrometer (a device that measures electric charge in a gas). The alpha particles could also be detected visually as they struck a zinc sulfide screen, causing an identifiable flash (scintillation). Geiger would build upon this technique in developing the electric radiation counter that bears his name.

Defines structure of the atom

In 1909 Rutherford instructed his Manchester colleague Marsden to study the scattering of alpha particles at large angles. Marsden observed that when alpha particles were fired at gold foil, a significant number of particles were deflected at unusually large angles; some particles were even reflected backward. Metals with a larger atomic number (such as lead) reflected back even more particles. It was not until late in 1910, however, that Rutherford assumed from this evidence the modern concept of the atom, which he announced early in 1911. He surmised that the atom did not resemble a "plum pudding" of positively charged nuclear particles with electrons embedded within like raisins, as had been suggested by Thomson. Instead, Rutherford suggested that the atom consisted of a very small, dense nucleus surrounded by orbiting electrons. Geiger and Marsden provided the mathematics to support the theory, and Bohr linked this concept with quantum theory (which states that energy is transmitted by radiation in the form of discrete units) to produce the model of the atom employed today.

After World War I began in 1914 Rutherford was called upon to serve on the British Navy's Board of Invention and Research. His main area of research for the board was devising a method for detecting German U-boats (submarines) at sea. His work established principles later applied in the development of sonar (a device that sends sound waves through water and picks them up after they strike some object and bounce back).

Succeeds in splitting the atom

Rutherford's team continued their work on alpha particle scattering during the war. Marsden observed in 1914 that alpha particles fired into hydrogen gas produced uncommon numbers of scintillations. Rutherford first concluded that the scintillations were being caused by hydrogen nuclei. However, he later observed the same effect when alpha particles were fired into nitrogen. In 1919, after a long series of experiments to exclude all possible explanations, Rutherford determined that the alpha particles were splitting the nitrogen atoms and that the extraneous (extra) hydrogen atoms were remnants of that split. Nitrogen was thus transmuted into another element. In 1925 English physicist Paul Maynard Stewart Blackett used the cloud chamber apparatus devised by Scottish physicist C. T. R. Wilson to verify Rutherford's observation and show that the atom had indeed split after it absorbed the alpha particle. (A cloud chamber is a vessel containing air saturated with water vapor whose sudden expansion reveals the passage of an ionizing particle by a trail of visible droplets.)

Directs the Cavendish Laboratory

In 1919 Rutherford was persuaded to succeed Thomson as director of the Cavendish Laboratory at Cambridge, a post he would hold until his death in 1937. Rutherford directed the Cavendish during its most fruitful research period in its history. English atomic physicist John D. Cockcroft and Irish experimental physicist Ernest Walton constructed the world's first particle accelerator there in 1932. Also known as an atom smasher, the particle accelerator increases the speed of charged subatomic particles or nuclei in order to demonstrate the transmutation of elements by artificial means. Also at Cavendish in the early 1930s English physicist James Chadwick confirmed the existence of the neutron, which Rutherford had predicted at least a decade earlier. Rutherford and Chadwick continued to bombard and split light elements with alpha particles. In 1934 Rutherford, with Marcus Oliphant and Paul Harteck, bombarded deuterium with deuterons (deuterium nuclei), achieving the first fusion (combination) reaction and production of tritium.

Becomes involved in politics

During this period Rutherford was actively involved in national and international politics as it related to science. He worked with the civilian Department of Scientific and Industrial Research (DSIR) to obtain grants for his scientific team and served as president of the British Association for the Advancement of Science from 1925 to 1930. Beginning in 1933 he served as president of the Academic Assistance Council, which was established to assist refugee Jewish scientists fleeing from Nazi Germany, where Jews were being persecuted. When the Soviet Union prevented Russian physicist Pyotr Kapitsa, a promising Cavendish scientist, from returning to Great Britain from the Soviet Union in 1934, Rutherford launched an ultimately futile effort to convince the Soviets to release him. Rutherford maintained close relations through correspondence with leading scientists in Europe, North America, Australia, and New Zealand. He corresponded with Bohr, German physicist **Max Planck,** German-American physicist **Albert Einstein** (see entries) and other theoreticians who were transforming physics through their works.

Receives awards and honors

Rutherford was bestowed with many honors during his long and distinguished career, most notably the Nobel Prize. He also received the prestigious Order of Merit, Britain's highest civilian honor, from King George V in 1925. Rutherford was knighted in 1914, and in 1931 he was made a nobleman (he was called Baron Rutherford of Nelson or Lord Rutherford). Rutherford died on October 19, 1937, in Cambridge. His cremated remains were buried near the graves of Isaac Newton and Charles Darwin at Westminster Abbey in London.

Further Reading

Andrade, E. N., da Costa, *Rutherford and the Nature of the Atom,* Doubleday, 1964.

Dictionary of Scientific Biography, Volume 12, Scribner, 1975.

Wilson, David, *Rutherford: Simple Genius,* MIT Press, 1983.

Albert Sabin

Born August 26, 1906
Bialystok, Russia (now Poland)
Died March 3, 1993
Washington, D.C.

Russian-born American virologist Albert Sabin developed the oral polio vaccine.

Noted virologist Albert Sabin developed an oral vaccine for polio that led to the virtual elimination in the Western Hemisphere of the once-dreaded disease. Sabin's long and distinguished research career included major contributions to virology (the branch of science that deals with viruses), including work that led to the development of live-virus vaccines. During World War II he developed effective vaccines against dengue fever and Japanese B encephalitis. The development of a live polio vaccine, however, was Sabin's major achievement.

Early life

Albert Bruce Sabin was born in Bialystok, Russia (now Poland), on August 26, 1906. His parents, Jacob and Tillie Sabin, immigrated to the United States in 1921 to escape extreme poverty. They settled in Paterson, New Jersey, where Sabin's father became involved in the silk and textile business.

After Sabin graduated from Paterson High School in 1923, one of his uncles offered to finance his college education if Sabin would agree to study dentistry. But during his dental education Sabin read Paul de Kruif's *Microbe Hunters* and was drawn to the science of virology, as well as to the romantic and heroic vision of conquering epidemic diseases.

Isolates B virus

After two years in the New York University (NYU) dental school, Sabin switched to medicine and promptly lost his uncle's financial support. He paid for school by working at odd jobs, primarily as a lab technician, and through scholarships. After earning a bachelor of science degree in 1928 he enrolled in the NYU College of Medicine. In medical school Sabin showed early promise as a researcher by developing a rapid and accurate system for typing (identifying) pneumococci, or the pneumonia bacteria. Sabin earned his medical degree in 1931, then served his residency at Bellevue Hospital in New York City, where he gained training in pathology (the study of disease), surgery, and internal medicine. In 1932, during his internship, Sabin isolated (separated from other substances for the purpose of studying) the B virus from a colleague who had died after being bitten by a monkey. Within two years Sabin had shown that the natural habitat of the B virus is the monkey and that it is related to the human herpes simplex virus. Completing his internship in 1934, Sabin conducted research for a year at the Lister Institute of Preventive Medicine in London, England.

Starts polio research

In 1935 Sabin returned to the United States and accepted a fellowship at the Rockefeller Institute for Medical Research. At the institute he resumed research of poliomyelitis (polio), a paralytic disease that had reached epidemic proportions in the United States at the time of Sabin's graduation from medical school. By the early 1950s polio had afflicted 13,500 out of every 100 million Americans. In 1950 alone more than 33,000 people had contracted polio. The majority of the victims were children.

Ironically, polio was once an endemic disease (a disease usually confined to a community, group, or region) spread by poor sanitation. As a result, most children who lived in households without indoor plumbing were exposed early to the virus. The vast majority of them did not develop symptoms and eventually became immune (protected) to later exposures. After the public health movement at the turn of the twentieth century began to improve sanitation, more families had indoor toilets. Therefore, children were not exposed at an early age to the virus and thus did not develop a natural immunity.

As a result, polio became an epidemic disease, spreading quickly through communities to other children who were not immune, regardless of race, creed, or social status. The disease was terrifying because it caused partial or full paralysis by lodging in the brain stem and spinal cord and attacking the central nervous system. Often victims of polio would lose complete control of their muscles and had to be kept in a respirator, later known as an iron lung, to help them breathe.

Studies effects of viruses

In 1936 Sabin and an associate, Peter K. Olitsky, used a test tube to grow some polio virus in the central nervous tissue of human embryos. While this was not a practical approach for developing the huge amounts of virus needed to produce a vaccine, the research nonetheless opened new avenues of investigation for other scientists. However, the scientists' discovery did reinforce the mistaken assumption that polio affected only nerve cells.

During World War II Sabin served in the United States Army Medical Corps. He was stationed in the Pacific theater, where he began investigations into insect-borne encephalitis (inflammation of the brain), sandfly fever, and dengue fever (an infectious disease transmitted by aëdes mosquitoes). He successfully developed a vaccine for dengue and conducted an intensive vaccination program on the island of Okinawa using a vaccine he had developed at Children's Hospital in Cincinnati, Ohio, that protected more than 65,000 military personnel against Japanese B encephalitis. Sabin eventually identified a number of antigenic (a substance, usually a protein or a carbohydrate, that stimulates a response from the immune system) types of sandfly fever and dengue viruses that led to the development of several attenuated (weakened, and therefore non-virulent) live-virus vaccines. A live-virus vaccine is a vaccine made of living virus that is diluted, or weakened, so that it triggers the immune system to fight off the disease without actually causing the disease itself.

IMPACT

Although Albert Sabin's polio vaccine was not the first, it eventually became the predominant mode of protection against polio throughout the Western world. In South America "Sabin Sundays" were held twice a year to eradicate the disease. The race to produce the first effective vaccine against polio was marked by intense and often hostile competition among scientists and their supporters. In addition to the primary goal of saving children, fame and fortune were at stake. Sabin, however, allowed his vaccine to be used free of charge by any reputable organizations as long as they met his strict standards in developing the appropriate strains.

Locates source of polio in human body

After the war Sabin returned to the University of Cincinnati College of Medicine, where he had previously accepted an appointment in 1937. With his new appointments as professor of research pediatrics and fellow of the Children's Hospital Research Foundation, Sabin resumed polio research. He and his colleagues began performing autopsies on everyone who had died from polio within a four-hundred-mile radius of Cincinnati. At the same time Sabin performed autopsies on monkeys. From these observations he found that the polio virus was present in humans in both the intestinal tract and the central nervous system. Sabin disproved the widely held belief that polio entered humans through the nose into the respiratory tract, showing that it first invaded the digestive tract before attacking nerve tissue. Sabin was also among the investigators who identified the three different strains of polio.

Sabin's discovery of polio in the digestive tract indicated that perhaps the polio virus could be grown in a test tube in nonnervous tissue as opposed to nerve tissue, which was costly and difficult to test. In 1949 scientists John Enders, Frederick Robbins, and Thomas Weller grew the first polio virus in human and monkey nonnervous tissue cultures, a feat that would earn them a Nobel Prize five years later. With the newfound ability to produce enough virus to conduct large-scale research efforts, virologists engaged in a race to develop an effective polio vaccine.

Competes with Salk to develop vaccine

At the same time that Sabin began his work to develop a polio vaccine, **Jonas Salk** (see entry), a young scientist at the University of Pittsburgh, entered the race. Both men were ambitious and committed to their own theory about which type of vaccine would work best against polio. While Salk concentrated on a killed-virus vaccine (in which the virus is made inactive with formalin, a formaldehyde-methanol solution), Sabin openly expressed his doubts about the safety of such a vaccine as well as its effectiveness in providing lasting protection. Sabin was convinced that an attenuated live-virus vaccine would provide safe, long-term protection.

In 1953 Salk seemed to have succeeded in producing a vaccine when he announced the development of a killed-virus vaccine. While many clamored for immediate mass field trials, Sabin, Enders, and others cautioned against mass inoculation until further effectiveness and safety studies were conducted. But Salk had won the support of the National Foundation for Infantile Paralysis, and in 1954 a massive field trial of the vaccine was held. A year later the vaccine was pronounced effective and safe. Just fourteen days after the announcement, however, five children in California contracted polio after taking the Salk vaccine. More cases began to occur, with 11 out of 204 people stricken and eventually dying. After the U.S. Surgeon General ordered a halt to the vaccinations, a live virus was found in certain batches of the manufactured vaccine. Eventually, manufacturing safeguards were mandated (ordered), and the Salk vaccine was again made available to the public. By 1961 the incidence of polio had been dramatically reduced in the United States.

Two polio patients in iron lungs converse at the Rancho Los Amigos Respiratory Center near Los Angeles, California, in August 1954.

Experiments with oral vaccine

Sabin turned his attention to developing a vaccine that, unlike the injected Salk vaccine, could be administered through the mouth. By orally administering the vaccine, Sabin wanted it to multiply in the intestinal tract. He tested individual virus particles on the central nervous system of monkeys to see whether the virus did any damage. According to various estimates, Sabin performed meticulous experiments on nine to fifteen thousand monkeys and hundreds of chimpanzees. Eventually he diluted three mutant strains of polio that seemed to stimulate antibody production in chimpanzees (antibodies counteract bacteria and other toxins). Sabin immediately tested the three strains on himself and his family, as well as research associates and volunteer prisoners from Chillicothe Penitentiary in Ohio.

Finally achieves success

Results of the tests showed that the viruses produced an immunity to polio with no harmful side effects. But by now the public and much of the scientific community were committed to the Salk vaccine. Two scientists working for Lederle Laboratories had also developed a live-virus vaccine. However, when the Lederle vaccine was tested in Northern Ireland in 1956, it proved dangerous, as it sometimes reverted to a virulent state. Although Sabin could not secure funding for a large-scale clinical trial in the United States, he remained undaunted. He was able to convince the Health Ministry in the Soviet Union to try his vaccine in massive trials to fight a polio epidemic that was claiming eighteen to twenty thousand victims a year in that country. By then Sabin was receiving the political backing of the World Health Organization in Geneva, Switzerland, which had previously been using the Salk vaccine to control the outbreak of polio around the world. Officials now believed that Sabin's approach would one day eradicate the disease.

Sabin began giving his vaccine to Russian children in 1957, inoculating millions over the next several years. Although the U.S. Public Health Service approved the manufacture of the vaccine in the United States, the agency would not permit its use. The Salk vaccine remained the vaccine of choice until a pediatrician in Phoenix, Arizona, named Richard Johns organized a Sabin vaccine drive. The vaccine was supplied free of charge, and many physicians provided their services without a fee. As the success of this effort spread, the Sabin vaccine soon became the preferred vaccine to ward off polio. By 1993 health organizations reported that polio was close to extinction in the Western Hemisphere.

Remains active in later life

Sabin married his first wife, Sylvia Tregillus, in 1935, and they had two daughters. After Sylvia Sabin died in 1966, he married Jane Warner; they later divorced. In 1972 he married Heloisa Dunshee De Abranches, a journalist. Sabin continued to work until he was nearly eighty. In 1980 he traveled to Brazil to help combat a new outbreak of polio. He antagonized Brazilian officials, however, by accusing the government of falsifying data concerning the serious threat that polio still presented in that country. He officially retired from the National Institutes of Health in 1986. Sabin continued to be outspoken, saying in 1992 that he doubted whether a vaccine against the human immunodeficiency virus (HIV) was feasible. He died from congestive heart failure on March 3, 1993.

Further Reading

Beale, John, *Lancet,* March 13, 1993, p. 685.

Great Events From History II, Salem Press, 1991, pp. 1522–26.

Robinson, Donald, *The Miracle Finders,* David McKay, 1976, pp. 41–47.

Schmeck, Harold M., Jr., "Albert Sabin, Polio Researcher, 86, Dies," *New York Times,* March 4, 1993, p. B8.

Shorter, Edward, *The Health Century,* Doubleday, 1987, pp. 60–70.

Florence R. Sabin

Born November 9, 1871
Central City, Colorado
Died October 3, 1953
Denver, Colorado

The "first lady of American science," Florence R. Sabin studied the lymphatic and immune systems and was a pioneer in promoting educational opportunities for women.

American anatomist Florence R. Sabin is best known for her studies of the origin of the lymphatic system and the response of the immune system to infections. In addition to being a dedicated researcher and teacher, Sabin also was an influential promoter of equal rights for women. Sabin became the first female full professor at Johns Hopkins School of Medicine and the first woman to be elected president of the American Association of Anatomists. Later in life, as a public health administrator, she made a great contribution to a number of Colorado communities.

Pursues studies in medicine

Florence Rena Sabin was born on November 9, 1871, in Central City, Colorado, to George Kimball Sabin, a mining engineer, and Serena Miner, a teacher. When Sabin was four, the family relocated to Denver, Colorado; three years later her mother died. Sabin, her sisters, and their father soon moved to

Lake Forest, Illinois, where they lived with an uncle. After graduating from a boarding school in Vermont, Sabin joined her older sister at Smith College in Massachusetts, where she became particularly interested in mathematics and science. In 1893 she earned a bachelor of science degree, then decided to pursue a career in medicine to demonstrate that an educated woman could be as good in any field as an educated man.

Begins career at Johns Hopkins

The new Johns Hopkins Medical School in Baltimore, Maryland, was one of the only medical schools at the time that would accept both men and women. A group of prominent local women had raised the necessary funds to open the school, and in return they insisted that women be allowed to be educated there—a radical idea at the time. Sabin began her studies at Johns Hopkins in 1896 and was strongly influenced by one of her professors (and, later, colleague), Franklin P. Mall, whose specialty was anatomy. (An anatomist is a person who studies the structure of organisms.) Having realized she preferred research and teaching to practicing medicine, Sabin took a position in the department of anatomy at Johns Hopkins after graduating in 1900. Thus Sabin began a long and productive career in a new field of research—the development of the human lymphatic system.

Studies lymphatic system

Sabin began her studies of the lymphatic system to settle the controversy over how the system developed. The lymphatic system runs throughout the body and is made up of lymph glands (or nodes) that carry infection-fighting white blood cells throughout the body. Most researchers at the time believed the vessels that made up the lymphatics formed independently from the vessels of the circulatory system, specifically the veins. A small minority (including Sabin) believed that the lymphatic vessels arose from the veins themselves, budding outward as continuous channels. Tests performed on pig embryos that were already so old that they could be con-

IMPACT

Remembered as "the first lady of American science," Florence R. Sabin made such biological breakthroughs as determining how the lymphatic system is formed and describing much about its structure that had not previously been known. Sabin developed new methods for studying the development of blood vessels and blood cells in embryos, and she applied her research to the fight against tuberculosis, a dreaded lung disease.

sidered adult proved inconclusive, so Sabin set out to study smaller, younger embryos.

Sabin, by studying pig embryos in which the lymphatic system was in its earliest stages of formation, demonstrated that lymphatics did indeed arise from veins, and that these sprouts connected with each other as they grew outward. Thus she determined that the lymphatic system eventually developed from existing vessels. Even after her results were confirmed by others, however, they remained controversial. Nonetheless, her book on the subject, *The Origin and Development of the Lymphatic System,* won an international prize in 1903, and she was appointed associate professor of anatomy at Johns Hopkins in 1905. Twelve years later she became the first woman to be awarded a full professorship at the medical school.

Contributes to fight against tuberculosis

Sabin's work on lymphatics led her to study the development of blood vessels and blood cells in embryos. She concentrated on white blood cells called monocytes, which attack infectious bacteria such as the organism that causes tuberculosis, an extremely contagious lung disease. Because tuberculosis was a dreaded health menace at the time that Sabin was doing her research, the National Tuberculosis Association awarded her with a research grant in 1924 to support her work on the body's immune system.

Promotes equal rights for women

Despite her scientific successes, Sabin's most cherished cause was the advancement of equal rights for women in education, employment, and society. Since she considered herself equal to her male colleagues, she frequently voiced her support for educational opportunities for women in the speeches

Florence R. Sabin in her laboratory in the late 1920s.

she made upon receiving awards and honorary degrees. She was also an active suffragist, campaigning for the right of women to vote.

Ironically, Sabin's career at Johns Hopkins drew to a close in 1925 after she was passed over for the position of professor of anatomy and head of the department, which her mentor Franklin Mall had held and which was given to one of her former students. She subsequently accepted a position at the Rockefeller Institute for Medical Research (now Rockefeller University) in New York City, where she continued her study of the role of monocytes and other white cells in the body's defense against tuberculosis and other infections. The discoveries that she and her colleagues made concerning the ways in which the immune system responded to tuberculosis led her to her final research project: the study of antibody formation.

Receives awards and honors

Upon her retirement in 1938 Sabin moved to Denver, Colorado, where she devoted the next fifteen years of her life to public health issues. As head of the "Sabin Committee" she fought for improved public health laws and construction of more health care facilities. She also held a variety of important posts in the Denver public health care system. One of her accomplishments was starting a chest X-ray campaign that helped cut the Denver death rate from tuberculosis by 50 percent in two years.

Among the many awards and honors bestowed upon Sabin was the Trudeau Medal from the National Tuberculosis Association, the Lasker Award of the American Public Health Association, and the dedication of the Florence R. Sabin Building for Research in Cellular Biology at the University of Colorado Medical Center. In 1924 she had been elected president of the American Association of Anatomists, and soon after she became the first woman elected to membership in the National Academy of Sciences. Sabin died of a heart attack on October 3, 1953, while rooting for her favorite team, the Brooklyn Dodgers, in a World Series game. After her death the state of Colorado erected a bronze statue of her in the National Statuary Hall in the Capitol in Washington, D.C., where each state is permitted to honor two of its most respected citizens.

Further Reading

Bluemel, Elinor, *Florence Sabin: Colorado Woman of the Century,* University of Colorado Press, 1959.

Kage, Judith, *The Life of Florence Sabin,* Twenty-First Century Books, 1993.

Kronstadt, Janet, *Florence Sabin,* Chelsea House, 1990.

Sicherman, Barbara, ed., *Notable American Women: The Modern Period,* Belknap Press, 1980.

Oliver Wolf Sacks

Born July 9, 1933
London, England

Author and neurologist Oliver Wolf Sacks has helped shed new light on the powers of the human mind through his books, essays, and lectures. The author of such well-known works as *Awakenings* and *The Man Who Mistook His Wife for a Hat,* Sacks uses case studies from his own medical practice to call for fundamental changes in the treatment of mental illnesses. His stories of courageous patients coping with handicaps have inspired popular movies, plays, and even an opera, but they speak more eloquently on their own as a testament to the hardiness of the human spirit. In an era when medical technology is valued over human interaction and overburdened physicians treat diseases instead of people, Sacks stands out as a rare and truly dedicated physician.

"What I did regret, and what many of the patients did as well, was that [L-dopa] was not available 10 or 20 years before, when they had not lost so many connections to the world."

Comes from family of physicians

Sacks was born in London, England, in 1933. Although he has never renounced his British citizenship, he has lived in

IMPACT

Oliver Wolf Sacks's book *Awakenings* served as the inspiration for playwright Harold Pinter's drama *A Kind of Alaska* and the Hollywood feature film *Awakenings*. The powerful and moving film centers on the "awakening" of a patient from a thirty-year-long coma. Sacks is portrayed as Dr. Malcolm Sayer by Robin Williams, and his notable patient, Leonard, by Robert De Niro. Sacks recalled in a *New York Times* interview that Williams perfectly captured his mannerisms and behavior in the role. "Robin ... had my gestural repertoire, my vocal and emotional style. It was like having a sudden younger twin." The physician added that later, to his dismay, people thought he was imitating Williams, not the other way around.

the United States since 1960. Both of his parents were doctors, and he grew up in an intellectually stimulating household. The expectation, of course, was that he would enter the medical profession someday. However, he was interested in many subjects besides biology. Sacks was only six years old when the dangers of World War II in London forced his parents to send him to a boarding school in the country. The school was run by a bitter headmaster who treated Sacks and his brother harshly. Apparently the boys never received delicacies mailed from London by their parents and instead were fed a diet of vegetables that were usually given to cattle. This experience left a profound mark on Sacks, who never again felt that he could trust institutions to take care of people in a humane way.

Becomes an unusual doctor

At the end of the war Sacks returned to London and immersed himself in his studies. He was particularly fond of chemistry but also read widely in philosophy, physics, the natural sciences, and biology. Words came easily to him, leading to some interesting experiences. Once he and a group of teenage friends boldly called upon prominent scientist Julian Huxley to talk about evolution. The surprised scientist received them graciously. Sacks later recalled, "I think the great man was both amused and impressed by such undersized, ink-stained, and sort of grimy children."

Sacks received a bachelor's degree and medical training at Queen's College of Oxford University in the late 1950s. In 1960 he left for America and two years later began a residency in neurology (the study of the nervous system) at the University of California at Los Angeles. Even then the hefty Englishman was a curiosity, as he insisted on taking a humanistic approach to his patients. In one unusual incident he gave a motorcycle ride to a multiple sclerosis victim (one who suffers from a disease marked by patches of hardened tissue in the brain or spinal cord, resulting in paralysis). Feeling that the rigid medical profession could harm patients more than help them, he planned to spend his career in the research laboratory instead of the hospital.

Finds his true vocation

That decision was short-lived. While serving a fellowship in New York, Sacks accidentally dropped hamburger in a centrifuge (a machine that separates substances) and lost an important vial of myelin (a fatty material which forms a sheath around certain nerve fibers) that had been carefully extracted from earthworms. He left research and returned to the practice of medicine, first in a clinic for the treatment of migraine headaches and later at Beth Abraham Hospital in the Bronx, a borough of New York.

Almost from the outset of his career as a physician Sacks enjoyed writing. He had always kept notebooks, recording his thoughts and observations during spare moments. His first book, *Migraine,* was based on his work with patients suffering from these debilitating headaches, which can cause violent pain, vomiting, and visual problems. Like his other works, *Migraine* goes beyond a mere clinical approach to the disease and seeks to explain the possible origins of the affliction and its impact on those who suffer from it.

Prescribes L-dopa

During his years at Beth Abraham Hospital Sacks became intrigued by a group of patients who seemed to be in a frozen, trancelike state. The patients had all contracted encephalitis lethargica (sleeping sickness) during an epidemic that broke out shortly after World War II. Though they seemed to recover from the early symptoms of the illness, they had all developed more distressing symptoms that made them unable to move or speak. Sacks had read research papers about the use of a new drug, L-dopa, for patients with Parkinson's disease (a disease of the nervous system that causes noticeable tremors and severe muscular rigidity). Sacks began to wonder what might happen to his strange, frozen patients if he administered L-dopa to them. In 1969 he decided to try the drug, and some of his patients were miraculously "awakened." Sacks kept detailed notes on his experiment, expecting to publish the results in medical journals. Instead, the story of his eighty sleeping sickness patients found its way into the popular book *Awakenings.*

In *Awakenings,* which was first published in 1973, Sacks recounts the case histories of the L-dopa-treated patients. He follows their miraculous return to functional living and their tragic battle with the drug's side effects. Unfortunately, many of the patients suffered from emotional upheaval and outright shock due to the combined effects of the drug and the reality of waking up in a completely different world.

Writes more books

The success of *Awakenings* established Sacks as an author whose "reports from the far boundaries of human experience" brought a new understanding to the amazing powers of the human psyche. His other books have expanded upon the themes of illness as a complex interaction of mind, body, and lifestyle—and the often creative means by which people overcome enormous afflictions. In the case of his 1984 work *A Leg to Stand On,* Sacks wrote about his own experience, describing how he struggled to regain the use of a leg he had injured in a fall while mountain climbing in Scandinavia.

One of Sacks's best-known books, the 1985 collection of case histories titled *The Man Who Mistook His Wife for a Hat,* is about patients he has seen in his practice. It was the basis for an opera produced in London in 1986 and a play that was performed in New York in 1995. This work and the more recent *An Anthropologist on Mars* both offer glimpses into the lives of people afflicted with various neurological disorders, including autism (a mental disease that results in a withdrawal from reality), Tourette's syndrome (a disorder causing uncontrollable ticks and outbursts), and even color-blindness. In 1989 Sacks wrote about the experiences of the deaf in *Seeing Voices: A Journey Into the World of the Deaf,* an extended essay on the human need for shared communication.

Continues treating patients

A bachelor who continues to practice medicine in the Bronx, Sacks admits he was not seeking fame when he began publishing books. Nevertheless, he is quite satisfied that his work has resulted in a better understanding of neurological illnesses and new insights into the workings of the mind. When Sacks is not treating his patients, lecturing at the Albert Einstein College of Medicine at Yeshiva University, or writing, he enjoys swimming. On certain occasions in the water, Sacks once said, "I'm transformed from my different, hesitant, geriatric, terrestrial form to a fluid, beautiful porpoise form."

Further Reading

Entertainment Weekly, March 10, 1995.

New York Times, May 24, 1984; June 19, 1985; January 25, 1986; February 7, 1995; February 14, 1995.

People, February 11, 1991.

Sacks, Oliver Wolf, *An Anthropologist on Mars,* Knopf, 1995.

Sacks, Oliver Wolf, *Awakenings,* Duckworth, 1973, Doubleday, 1974, new edition, Summit Books, 1987.

Sacks, Oliver Wolf, *A Leg to Stand On,* Summit Books, 1984.

Sacks, Oliver Wolf, *The Man Who Mistook His Wife for a Hat, and Other Clinical Tales,* Duckworth, 1985, Summit Books, 1986.

Sacks, Oliver Wolf, *Seeing Voices: A Journey Into the World of the Deaf,* University of California Press, 1989.

Abdus Salam

Born January 29, 1926
Jhang, Pakistan

Pakistani physicist Abdus Salam shared the 1979 Nobel Prize for his work on electroweak force.

In 1968 Pakistani physicist Abdus Salam published a theory showing how the electromagnetic force and the weak nuclear force may be considered as separate and distinct functions of the single electroweak force. Experiments conducted at the European Center for Nuclear Research (CERN) in 1973 provided the evidence needed to substantiate Salam's theory. For this work he shared the 1979 Nobel Prize in physics with Sheldon L. Glashow and Steven Weinberg (see box), who worked independently on aspects of the electroweak force. Because of his personal experiences as a scientist in a Third World (developing) country that did not emphasize scientific research, Salam pushed for the establishment of the International Center for Theoretical Physics (ICTP) to provide Third World physicists the kind of instruction that is generally not available in their own homelands.

Becomes professor in London

Abdus Salam was born on January 29, 1926, in the small rural town of Jhang, Pakistan, to Hajira and Muhammed Hussain. At the age of sixteen he entered the Government College at Punjab University in Lahore, and in 1946 Salam received his master's degree in mathematics. He was then awarded a scholarship that allowed him to enroll at St. John's College at Cambridge University in England. In 1949 he earned a bachelor's degree in mathematics and physics, with highest honors. After spending a short time in his native Pakistan, where he learned there were no opportunities for him conduct research because he was the only theoretical physicist in the entire nation, Salam returned to Cambridge and received a Ph.D. in theoretical physics in 1952. He taught mathematics for two years at Cambridge and, in 1957, was appointed professor of theoretical physics at the Imperial College of Science and Technology in London, where he continues to hold this position.

Attacks the problem of force unification

Beginning in the mid-1950s Salam turned his attention to one of the fundamental questions of modern physics, the unification of forces. Scientists recognize that there are four fundamental forces governing nature—gravity, the strong nuclear force, the weak nuclear force, and electromagnetism—and that all four may be manifestations of a single basic force. By the early 1960s efforts toward unification were directed specifically at finding ways of relating the electromagnetic and weak forces. Electromagnetism causes such phenomena as sunlight and radio waves, and the weak force operates over very short distances within the nucleus and is responsible for some forms of radioactive decay.

Since the unity of forces would not actually be observable except at energy levels that currently exist only in cosmic radiation and in the most powerful of particle accelerators,

IMPACT

Abdus Salam studied one of the fundamental questions of modern physics, the unification of the forces governing nature. Since the actual unification of the forces would not be observable except at extraordinary energy levels, attempts to prove unification theories are mostly theoretical exercises involving esoteric mathematical formulations. In the 1960s Salam derived a mathematical theory that unifies two of the four basic forces, electromagnetism and weak nuclear force.

attempts to prove unification theories are, to some extent, theoretical exercises that may be testable initially only by checks of internal consistency and only later by experimental studies. During the 1960s Salam and physicist Steven Weinberg formulated a mathematical theory that unifies the electromagnetic and weak forces into the electroweak force. They produced the same theory starting from two very different origins and following two different lines of reasoning.

One of the predictions resulting from the new electroweak theory was the existence of previously unknown weak "neutral currents," as anticipated by Salam and Weinberg. These currents were first observed in 1973 during experiments conducted at the CERN in Geneva and later at the Fermi National Accelerator Laboratory in Batavia, Illinois. A second prediction, the existence of force-carrying particles, was verified in a later series of experiments also carried out at CERN in 1983. By that time, Salam, Glashow, and Weinberg had been honored for their contributions to the electroweak theory with the 1979 Nobel Prize in physics.

Concerned with the Third World

Salam's concern over the lack of science opportunities in Pakistan and other developing nations prompted him to help establish the International Center for Theoretical Physics (ICTP) in Trieste, Italy, in 1964. He has served as director of the institute since its founding. The ICTP invites outstanding theoretical physicists to teach and lecture aspiring students on their own areas of expertise. Additionally, in Pakistan Salam served as a member of the Atomic Energy Commission (1958–74) and the Science Council (1963–75), as Chief Scientific Advisor to the President (1961–74), and as chairperson of the

Steven Weinberg and Sheldon L. Glashow, American Physicists

Abdus Salam shared the 1979 Nobel Prize in physics with American physicists Steven Weinberg (1933–) and Sheldon L. Glashow (1932–) for their contributions toward the development of a theory unifying the electromagnetic and weak nuclear forces. In the 1960s Salam and Weinberg independently formulated a mathematical theory that unifies the electromagnetic and weak forces into the electroweak force. Further, both Weinberg and Salam predicted that one class (the leptons, containing electrons and neutrinos) of the three particles inherent in the weak force could be found in "neutral currents," a current in which no electronic charge is exchanged. In 1971 Weinberg had indicated that collisions of neutrinos (a particle having neither mass nor charge) and matter produced a neutral current that contained one of the three weak force-carrying particles. In 1973 experiments at the European Center for Nuclear Research (CERN) confirmed the existence of these currents, and a decade later the three types of particles were detected in further experiments carried out at CERN by a group headed by Italian and Dutch physicists.

Shortly after the Salam-Weinberg theory was announced, Glashow found a method for extending the theory to the other elementary particles, called quarks (fundamental subatomic particles found in protons). In order to do so, he found it necessary to invent a new property (attribute) for particles, which he designated as "charm." Glashow's prediction was soon confirmed by research at the CERN as well. Further confirmation was obtained in 1974 with the discovery of the J/Y particle, which is composed of a charm quark and its antiparticle, the anticharm quark. These discoveries earned Salam, Weinberg, and Glashow the Nobel Prize and may have brought scientists closer to a unified theory of the universe.

Space and Upper Atmosphere Committee. He has been involved in numerous international scientific activities as well as a variety of governmental agencies. He was a member and chairman of the United Nations Advisory Committee on Science and Technology, vice president of the International Union of Pure and Applied Physics, and a member of the Scientific Council of the Stockholm International Peace Research Institute. Besides receiving the Nobel Prize in physics, Salam has been awarded more than two dozen honorary doctorates and has received more than a dozen major awards, including the

Sheldon L. Glashow, Abdus Salam, and Steven Weinberg at the 1979 Nobel Prize ceremonies in Stockholm, Sweden.

Atoms for Peace Award for 1968, the Royal Medal of the Royal Society, the John Torrence Tate Medal of the American Institute of Physics, and the Lomonosov Gold Medal of the U.S.S.R. Academy of Sciences.

Further Reading

Hall, Nina, "A Unifying Force for Third World Science," *New Scientist,* January 27, 1990, p. 31.

"Physics for the Poor," *The Economist,* November 18, 1989.

Powell, John, "What the Third World Can Handle," *Bulletin of the Atomic Scientists,* May 1989.

Salam, Abdus, *Ideas and Realities: Selected Essays of Abdus Salam,* World Scientific, 1987.

Siddique, A. K., Abdus Salam, and others, "Why Treatment Centers Failed to Prevent Cholera Deaths Among Rwandan Refugees in Goma, Zaire," *The Lancet,* February 11, 1995.

Wasson, Tyler, ed., *Nobel Prize Winners,* H. W. Wilson, 1987, pp. 914–16.

Weber, Robert L., *Pioneers of Science: Nobel Prize Winners in Physics,* American Institute of Physics, 1980, pp. 263–64.

Jonas Salk

Born October 28, 1914
East Harlem, New York
Died June 23, 1995
La Jolla, California

"I don't want to go from one crest to another. To a scientist, fame is neither an end nor even a means to an end."

American microbiologist Jonas Salk was celebrated worldwide for developing the vaccine that ended the threat of the dreaded polio virus. He discovered that a "killed virus," or a virus made inactive with a formaldehyde solution, is capable of serving as an antigen, or an agent that prompts the body's immune system to produce antibodies that will attack invading organisms. This realization enabled Salk to develop a polio vaccine composed of killed polio viruses, producing the necessary antibodies to help the body ward off the disease without itself inducing polio.

Chooses medical career

The oldest son of Jewish immigrants from Poland, Jonas Edward Salk was born in East Harlem, New York, on October 28, 1914. His father, Daniel B. Salk, was a garment worker, and his mother, Dora Press Salk, encouraged his academic talents. They sent Salk to Townsend Harris High School for the gifted, where he showed he was both highly motivated and high achieving. After graduating at the age of fifteen he enrolled in the legal program at the City College of New York. Ever curious, Salk also attended science courses and quickly decided to switch fields. He earned a bachelor's degree in science in 1933, at the age of nineteen, and went on to the New York University School of Medicine. Initially he scraped by on money his parents had borrowed for him, but after his first year he received scholarships and fellowships. During his senior year he began working with Thomas Francis Jr., with whom Salk would collaborate on some of the most important work of his career.

The day after Salk was awarded his medical degree in 1939 he married Donna Lindsay, a social worker. They later had three sons: Peter, Darrell, and Jonathan. In 1939 Salk also began a two-year internship at Mount Sinai Hospital in New York. Upon completing his internship Salk accepted a National Research Council fellowship and moved to the University of Michigan in Ann Arbor, where he joined Francis, who had become head of the department of epidemiology the previous year. Working on a project for the U.S. Army, the team strove to develop a flu vaccine. Their goal was a "killed-virus" vaccine, in which the virus is made inactive (killed) by formalin, a formaldehyde-methanol solution. The killed virus would be able to kill the live flu viruses in the body as well as produce antibodies that could fight off future invaders of the same type, thus resulting in immunity to the flu. By 1943 Salk and Francis had developed a killed-virus vaccine that proved effective against both type A and B influenza viruses. Salk and Francis were now in a position to begin clinical trials.

Rejuvenates viral research laboratory

In 1946 Salk was appointed assistant professor of epidemiology at Michigan. Around this time he extended his research to cover not only viruses and the body's reaction to them but also their epidemic effects in populations. The following year he accepted the position of associate research professor at the Virus Research Laboratory at the University of

IMPACT

The Salk vaccine was not the first vaccine against polio, but it was the first to be found safe and effective. By 1961 polio cases in the United States had been reduced by 96 percent. During the 1960s the Sabin live-virus vaccine, developed by **Albert Sabin** (see entry), began to replace the Salk vaccine because of its effective one-time oral administration. However, the Salk vaccine is still considered a triumph of medical science.

Pittsburgh School of Medicine. When Salk arrived, however, he found the staff had no experience with the kind of basic research he conducted. In order to bring the lab up to par, he solicited financial support from outside benefactors. Soon the Virus Research Laboratory represented the cutting edge of viral research.

Begins work on polio vaccine

In addition to building a respectable laboratory, Salk also devoted a considerable amount of time to writing scientific papers on a number of topics, including the virus that causes poliomyelitis (known simply as polio). His work came to the attention of Daniel Basil O'Connor, the director of the National Foundation for Infantile Paralysis, an organization that had long been involved in the treatment and rehabilitation of polio victims. When the two men met, O'Connor was so impressed with Salk that he pledged nearly the entire budget of the foundation to Salk's vaccine research efforts. Polio, which can be traced back to ancient Egypt, causes permanent paralysis or chronic shortness of breath in those it strikes, often leading to death. Children are especially vulnerable to the virus. The University of Pittsburgh was one of four universities engaged in trying to sort and classify the more than one hundred known varieties of polio virus.

By 1951 Salk was able to assert with certainty that all polio viruses fell into one of three types, each having various strains. Some of the viruses were highly infectious, while others were barely so. Once he had determined the virus types, Salk began to work on a vaccine. His first challenge was to obtain enough of the virus to be able to develop a vaccine in doses large enough to have an impact. This task was particularly difficult since viruses, unlike culture-grown bacteria (bacteria produced in a laboratory), need living cells to grow. A breakthrough came when the team of John F. Enders (see box), Thomas Weller, and Frederick Robbins found that the polio virus could be grown in embryonic tissue. For their discovery Enders, Weller, and Robbins were awarded a Nobel Prize in 1954.

Salk subsequently grew samples of all three varieties of polio virus in cultures of monkey kidney tissue, then killed the virus with formaldehyde. He believed it was essential to use a killed virus rather than a live virus (which is weak, though still live), because the live-virus vaccine would have a much greater chance of accidentally inducing polio in inoculated children. He therefore exposed the viruses to formaldehyde for nearly thirteen days. Although Salk could detect no presence of viruses in the sample after only three days, he wanted to establish a wide safety margin. After an additional ten days of exposure, he reckoned that there was only a one-in-a-trillion chance of there being a live virus particle in a single dose of his vaccine. Before proceeding to human clinical trials, Salk tested the vaccine on monkeys. The results were successful.

Encounters skepticism

Despite Salk's confidence in his findings, many of his colleagues were skeptical, believing that a killed-virus vaccine could not possibly be effective. His dubious standing was further compounded by the fact that he was relatively new to polio vaccine research. His competitors in the race to develop the vaccine—most notably **Albert Sabin** (see entry), the leading proponent of a live-virus vaccine—had been working for years and were irked by the presence of this upstart with unorthodox ideas.

As the field narrowed the division between the killed-virus and the live-virus camps widened, and what had once been a polite difference of opinion became a serious ideological conflict. Salk and his primary backer, the National Foundation for Infantile Paralysis, had received minimal support. Undeterred, Salk continued his research. To test the strength of his vaccine, in 1952 Salk administered a type I vaccine to children who had already been infected with the polio virus.

Afterward he measured their antibody levels. The results clearly indicated that the vaccine produced large amounts of antibodies. Buoyed by this success, Salk extended the clinical trial to include children who had never had polio.

Conducts largest medical experiment in U.S. history

In May 1952 Salk initiated preparations for a massive field trial in which over four hundred thousand children would be vaccinated. The largest medical experiment that had ever been carried out in the United States, the test began in April 1954. It was directed by Francis and sponsored by the National Foundation for Infantile Paralysis. More than one million children between the ages of six and nine took part in the trial, each receiving a button that proclaimed them a "Polio Pioneer." One-third of the children were given doses of the vaccine consisting of three injections—one for each of the types of polio virus—plus a booster shot. A control group of the same number of children was given a placebo (an inert substance), and a third group was given nothing.

At the beginning of 1953, while the trial was still at an early stage, Salk's encouraging results were made public in the *Journal of the American Medical Association.* Predictably, media and public interest were intense. Anxious to prevent sensationalized accounts of his work, Salk agreed to comment on the results thus far during a scheduled radio and press appearance. Since this arrangement did not conform to accepted scientific protocol for making such announcements, some of Salk's fellow scientists accused him of being a publicity seeker. Saying he had been motivated only by the highest principles, Salk was deeply hurt.

Becomes a reluctant celebrity

Despite the doomsayers, on April 12, 1955, the vaccine was officially pronounced effective, potent, and safe in almost 90 percent of cases. The meeting at which the announcement was made was attended by five hundred of the world's top sci-

John F. Enders, American Virologist

American virologist John F. Enders's (1897–1985) research on viruses and his advances in tissue culture enabled microbiologists Jonas Salk and Albert Sabin to develop their vaccines against polio, a major crippler of children in the first half of the twentieth century. In 1946 Enders established the Infectious Disease Laboratory at Boston Children's Hospital, and by 1948 he had two assistants, Frederick Robbins and Thomas Weller. Although Enders and his colleagues did their research primarily on measles, mumps, and chicken pox, their lab was partially funded by the National Foundation for Infantile Paralysis, an organization established to help victims of polio and to find a vaccine or cure for the disease.

Enders's revolutionary discovery concerning polio occurred almost accidentally. During an experiment on chicken pox, Weller produced too many cultures of human embryonic tissue. So as not to let them go to waste, Enders suggested putting polio viruses in the growing cultures. To his and his colleagues' surprise, the virus began growing in the test tubes. The publication of these results in a 1949 *Science* magazine article caused major excitement in the medical community, because experiments in the 1930s had indicated that the polio virus could grow only in nervous system tissue. The technique of cultivating the virus and studying its effects represented a new development in viral research.

Although Enders could have developed a vaccine for polio, he did not take the opportunity. The exact reason for his decision is unclear; some suggest that Enders was reluctant to submit himself to the restrictions that the National Foundation might have placed on his research. Nonetheless, because his breakthrough made it possible to develop a vaccine against polio, Enders, Robbins, and Weller were awarded the Nobel Prize for medicine or physiology in 1954. Interestingly, Enders initially opposed Salk's proposal to vaccinate against polio by injecting killed viruses into an uninfected person to produce immunity. He feared this procedure would actually weaken the immunity of the general population by interfering with the ways the disease developed. In spite of their disagreements, Salk expressed gratitude to Enders by stating that he could not have developed his vaccine without the help of Enders's discoveries.

entists and doctors, one hundred fifty journalists, and sixteen television and movie crews. The success of the trial catapulted Salk to instant stardom. He was inundated with offers from Hollywood and with invitations from manufacturers to endorse their products. Salk received a citation from President Dwight D. Eisenhower and addressed the nation from the White House Rose Garden. He was also awarded a congressional medal for great achievement in the field of medicine. Yet when he was nominated for a Nobel Prize, contrary to popular expectation he did not receive the award. Nor was he accepted for membership in the National Academy of Sciences. This slight most likely was a reflection of the discomfort the scientific community still felt about the level of publicity Salk attracted and of the continued disagreement over his methods. Wishing to escape the glare of the limelight, he tried to retreat into his laboratory.

Vaccination campaign temporarily halted

Unfortunately, tragedy served to keep the attention of the world media focused on Salk. Just two weeks after he announced his discovery, eleven of the children who had received the vaccine developed polio. Several more cases soon followed. Altogether, about two hundred children developed paralytic polio, eleven fatally. On May 7 the Surgeon General called the vaccination campaign to a halt. A thorough investigation determined that the defective vaccines were found to have originated from the same source, Cutter Laboratories in California. Cutter had used faulty batches of virus culture that were resistant to formaldehyde. After furious debate and the adoption of standards that would prevent such a reoccurrence, inoculation resumed. By the end of 1955 seven million children had received their shots, and over the course of the next two years more than two hundred million doses of the Salk vaccine were administered, without a single instance of vaccine-induced paralysis. By mid-1961 there had been a 96 percent reduction in the number of cases of polio in the United States compared to the five-year period prior to the vaccination campaign.

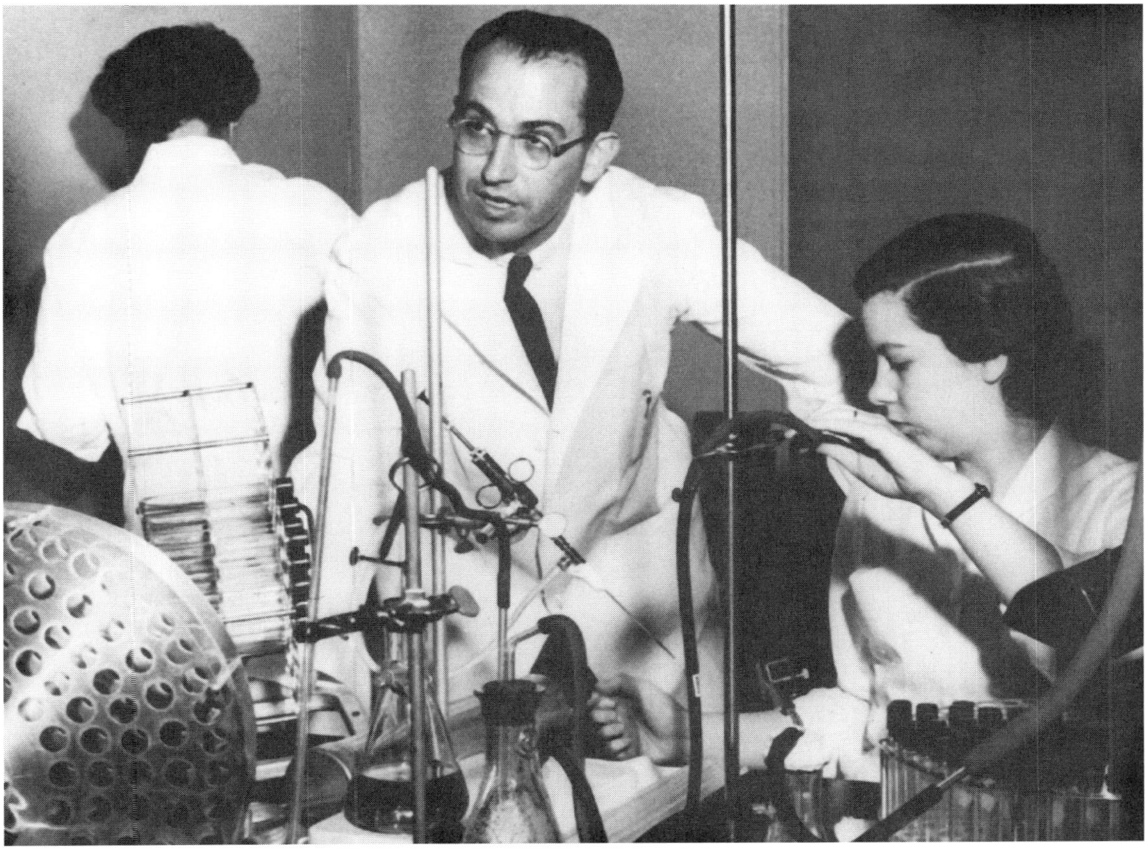

Jonas Salk oversees an assistant's work during a step in the production of the polio vaccine in March 1954.

After the initial inoculation period ended in 1958, Salk's killed-virus vaccine was replaced by a live-virus vaccine developed by Sabin. Use of the Sabin vaccine was advantageous because it could be administered orally rather than intravenously, and it required fewer "booster" inoculations. However, Salk remains known as the man who defeated polio.

Founds Institute for Biological Studies

In 1954 Salk was appointed professor of preventative medicine at the University of Pittsburgh, then in 1957 he became professor of experimental medicine. The following year he began work on a vaccine to immunize against all viral diseases of the central nervous system. As part of this research Salk performed studies of normal and malignant cells, which pertained to problems encountered in cancer research. In 1960 he founded the Salk Institute for Biological Studies in La Jolla, California. Heavily funded by the National Foundation for Infantile Paralysis (by then known as the March of Dimes), the institute attracted some of the best scientists in the world, all drawn by Salk's promise of full-time, uninterrupted biological research.

When his new institute opened in 1963, Salk became the director and devoted himself to the study of multiple sclerosis and cancer. He made headlines again in 1967 when he married Frances Gilot, the first wife of the Spanish artist Pablo Picasso. During the 1970s Salk turned to writing, producing books about the philosophy of science and its social role. In 1977 he received the Presidential Medal of Freedom. In the early 1990s many people looked to him as the scientist would might finally develop a vaccine against the human immunodeficiency (HIV) virus. Although Salk continued to strive for scientific breakthroughs, he chose not to move into this field of research. Salk died of heart failure in La Jolla, California, on June 23, 1995.

Further Reading

Curson, Marjorie, *Jonas Salk,* Silver Burdett, 1990.

Hargrove, Jim, *The Story of Jonas Salk and the Discovery of the Polio Vaccine,* Children's Press, 1990.

Robinson, Donald, *The Miracle Finders,* David McKay, 1976.

Rowland, Donald, *The Polio Man: The Story of Dr. Jonas Salk,* Roy Publishing, 1961.

Siemens, Pliny, *The Great Scientists,* Grolier, 1989, p. 137.

Berta Scharrer

Born December 1, 1906
Munich, Germany
Died in 1995

German-born American biologist Berta Scharrer and her biologist husband Ernst Scharrer pioneered the field of neuroendocrinology, the study of the interaction between the nervous system and the endocrine glands and their secretions. Fighting against accepted scientific beliefs about cells, as well as against prejudice toward women in the sciences, Scharrer established the concept of neurosecretion, or the releasing of substances, such as hormones, by nerve cells. A respected educator, she was also instrumental in founding the department of anatomy at the Albert Einstein College of Medicine in New York City. Scharrer is best known, however, for her research on the nerve cells of the South American cockroach brain, which gave her insights into the functioning of the human nervous system.

"Prospects for an academic career in Germany were bleak and, for a woman, virtually nonexistent."

Develops interest in science

Berta Scharrer was born in Munich, Germany, on December 1, 1906, to Karl Phillip and Johanna (Greis) Vogel.

IMPACT

Berta Scharrer, along with her husband Ernst, was a pioneer in the field of neuroendocrinology, the study of the interaction between the nervous system and the endocrine glands (which produce hormones) and their secretions. They promoted their theory of neurosecretion—that a class of nerve cells releases substances such as hormones, which prompt chemical reactions—even when such an idea was considered unorthodox by the scientific community. After another German scientist made it easier to study neurons, neurosecretion not only became an accepted fact, but the cornerstone of the emerging field of neuroendocrinology.

She developed an early interest in science and attended the University of Munich, earning her Ph.D. in biology in 1930. For her thesis on the correlation between sweetness and nutrition in various sugars, she conducted experiments on honeybees under the direction of Nobel Prize-winning zoologist **Karl von Frisch** (see entry). Upon graduation Scharrer took a position as a research associate at the Research Institute of Psychiatry in Munich. In 1934 she married Ernst Scharrer, a biologist whom she had met at the university.

Introduces theory of neurosecretion

The Scharrers spent the first three years of their marriage in Munich before moving to Frankfurt, where Ernst became director of the Edinger Institute for Brain Research. At this time they started a collaboration that lasted more than thirty years, until Ernst's death in 1965. They began working on a theory Ernst had formulated in 1928, when he had discovered what he termed nerve-gland cells (which later became known as neurosecretory cells) in a species of fish. He had also suggested the rather startling hypothesis that some nerve cells, like cells of the endocrine system, actually were involved in secreting hormonal substances. However, his thesis was not accepted by the more conservative members of the scientific community because the idea that neurons (cells that are the fundamental functional units of nervous tissue) had a dual function was considered implausible.

According to scientists at that time, either cells secreted hormones, in which case they were endocrine cells belonging to the endocrine system, or they conducted electrical impulses, making them nerve cells belonging to the nervous system. In the course of their research, however, the Scharrers found a distinct class of cells that performed both functions. They discovered that nerve-gland cells actually serve as a channel between the nervous system and the endocrine system. Some of the neurohormones (hormones secreted by or acting on a part of the nervous system) secreted by neurosecretory cells actually control the release of other hormones through the pituitary gland. (The pituitary is a major endocrine gland located just below the brain. Controlled by the adjacent hypothalamus gland, the pituitary controls other endocrine glands.)

The Scharrers set out to prove their concept of neurosecretion. Organizing their research, they divided up the animal kingdom between them: Berta took the invertebrates (organisms without spines), and Ernst took the vertebrates (organisms with spines). While working as a research associate at the Neurological Institute of the University of Frankfurt, Berta discovered nerve-gland cells in mollusks (marine invertebrates) in 1935, in worms in 1936, and in insects beginning in 1937. The Scharrers continued their work until Adolf Hitler's dictatorship made life in Germany unbearable. Under Hitler, the National Democratic Socialist (Nazi) regime began to enact repressive laws against the German Jews, eventually executing millions of them in what became known as the Holocaust. Even though the Scharrers weren't Jewish, they were appalled by the government's actions. As a result, they immigrated into the United States in 1937.

Begins research with cockroaches

The Scharrers traveled to America via the Pacific Ocean, collecting specimens for research along the way. Upon their arrival in the United States, they joined the anatomy department at the University of Chicago for a year. At Chicago Berta Scharrer began the research on cockroaches that would later make her famous. Having no real professional standing and therefore lacking a budget for lab animals, she reportedly collected the roaches in the basement of the lab and used them for experiments. In 1938 the couple moved to New York City, where Scharrer continued her insect research while her husband served for two years as a visiting investigator at the

Rockefeller Institute. During this time she began experimenting on South American cockroaches. Apparently a shipment of lab monkeys had arrived one day from South America, and Scharrer saw cockroaches—*Leucophaea maderae*—scurrying around in the bottom of the monkey cage. She immediately retrieved them and found that they made better research subjects because they were slower than the American cockroach. From that point forward Scharrer used the South American roaches, which traveled with her wherever she and her husband moved.

Scharrers prove theory of neurosecretion

By 1940 the Scharrers were ready to present the results of their research at the annual meeting of the Association for Research in Nervous and Mental Diseases. This was the first presentation on the concept of neurosecretion in the United States, and it was warmly received. That same year Ernst took a position as assistant professor in the anatomy department of Western Reserve University School of Medicine in Cleveland, Ohio. Scharrer was offered a fellowship in the histology (the study of the microscopic structure of animal and plant tissue) laboratory, which gave her research facilities but, again, she had limited professional standing and received no salary. According to some accounts, she was permitted to attend department seminars only if she made tea for the faculty. During these years Scharrer accomplished some of her most important research on the location of neurosecretory cells and their role in animal development using her South American cockroaches.

Theory becomes accepted fact

After World War II Scharrer won a Guggenheim Fellowship that allowed her to continue her research. She also became an assistant professor in 1947 at the University of Colorado Medical School in Denver, where her husband had also accepted a position. The next few years were a productive period for the Scharrers, as the theory of neurosecretion was

Susan E. Leeman, American Neuroendocrinologist

American neuroendocrinologist Susan E. Leeman (1930–) is known for her work with substance P and neurotensin, peptides that help govern the functioning of the nervous, endocrine, and immune systems. (Peptides are compounds containing two or more amino acids, the building blocks of proteins.) While she was trying to purify corticotropin (a protein-hormone preparation used in the treatment of rheumatoid arthritis and rheumatic fever), Leeman found a peptide that could stimulate the secretion of saliva. The chemical turned out to be substance P, which had been discovered in the 1930s but had never been isolated (separated from other substances for individual study).

Leeman and her colleagues, working at Brandeis University in Massachusetts, isolated and characterized the peptide. A nerve transmitter that has many functions in the body, substance P is distributed throughout both the central and peripheral nervous systems and the spinal cord. Substance P is important in the interaction between the nervous system and the immune system and seems to play a role in inflammation in the body. During her work with substance P, Leeman discovered another peptide, neurotensin, which is involved in relaxation and contraction of the blood vessels. She and her colleagues isolated neurotensin and determined its amino acid sequence. Found in both the central nervous system and the gastrointestinal tract, neurotensin may be involved in psychiatric disorders and, possibly, regulation of the menstrual cycle. Leeman was the first woman elected to the Physiology and Pharmacology section of the National Academy of Sciences.

beginning to gain acceptance around the world after a German scientist was able to successfully stain neurosecretory granules (the packaging for neurohormones that some neurons secrete). This scientist's staining made it possible to study the fine structure of such granules and follow their course upon secretion. As a result, neurosecretion not only became an accepted fact but the cornerstone of the emerging field of neuroendocrinology.

By 1950 the transmission of chemicals in the body was recognized as taking place at the synapse (the point at which a nervous impulse passes from one neuron to another) in conjunction with the release of an electrical charge (nervous impulse). These advancements not only confirmed the Scharrers' work, but also paved the way for progress in their research. Scharrer applied the new findings to her own experiments on the maturation of the ovarian systems of her cockroaches. She obtained results that verified earlier findings in the endocrinology of invertebrates.

Makes important contributions

In 1955 the Scharrers were offered joint positions at the new Albert Einstein College of Medicine at Yeshiva University in New York City, where Ernst was appointed head of the anatomy department. Berta was named a full professor in the same department, thus for the first time gaining professional recognition. She taught histology and continued her research on insect glands. Using the electron microscope Scharrer produced some of the earliest descriptions of the insect nervous system and the neurosecretory system.

In 1963 the Scharrers published *Neuroendocrinology,* one of the basic texts in the new discipline. Two years later Ernst died in a swimming accident in Florida. Despite her husband's death, Scharrer carried on with their research, acting as chairperson of the anatomy department for two years until a successor could be found. She also went on to describe the fine structure of the neurosecretory cell. Scharrer illustrated that the neurosecretory cell, like other neurons (nerve cells), is composed of a cell body, projecting dendrites (branches that conduct impulses toward the body of a nerve cell), the extending long axon (which conducts impulses away from the cell body), and synapse contacts at one end. She also showed that neurosecretory cells have special fibers, allowing for feedback, as well as neurohemal organs (points at which neurohormones pass into the blood stream).

Scharrer's research demonstrated that, like other hormones, neurosecretory cells can affect targets near to them or at a distance from them through the blood stream. Scharrer also investigated the make-up of the material secreted by the cells, discovering that it is a peptide or polypeptide. A peptide is a compound containing two or more amino acids (the chief components of proteins) linked through the amino group of one acid and the carboxyl group of the other. A polypeptide, also called a protein, is a peptide consisting of many units of amino acids.

Receives awards and honors

In addition to conducting research and teaching, Scharrer was coeditor of the journal *Cell and Tissue Research.* Scharrer became an emeritus professor of anatomy and neuroscience at Albert Einstein College of Medicine in 1978. She was honored with a National Medal of Science in 1983 for her "pioneering contributions in establishing the concept of neurosecretion and neuropeptides in the integration of animal function and development." She also won the F. C. Koch Award of the Endocrine Society in 1980, the Henry Gray Award of the American Association of Anatomists in 1982, and was honored by her native country with the Kraepelin Gold Medal from the Max Planck Institute in Munich in 1978 and the Schleiden Medal in 1983. She was a member of the National Academy of Sciences and held honorary degrees from colleges and universities around the world, including Harvard and Northwestern Universities.

Continues research in final years

During the final years of her life Scharrer's research contributed to the new field of psychoneuroimmunology, the study of how psychological and emotional states can affect the immune system. One of her last projects was studying natural opiates (substances that cause relaxation) in the brain with George Stefano, a neuroscientist at the State University of New York at Old Westbury, Long Island. She and Stefano succeeded in isolating a receptor (a cell that receives stimuli) that responds only to morphine secreted by human nerve cells. Stefano has used this research to conduct studies of the effects of morphine on hospital patients. According to his preliminary findings, when a patient is given a large dose of morphine before surgery the drug reduces inflammation in the body and helps the patient recover more quickly.

Scharrer closed her lab in late 1994, a year before her death, so she could spend more time writing and assisting other researchers with their experiments. She shipped her South America cockroaches to Stefano at Old Westbury, where they are still being used for research on neurosecretion. A species of cockroach, the *Scharrerae,* has been named in her honor.

Further Reading

"Insects as Models of Neuroendocrine Research," *Annual Review of Entomology,* Volume 32, 1987, pp. 1–16.

"Neurosecretion: Beginnings and New Directions in Neuropeptide Research," *Annual Review of Neuroscience,* Volume 10, 1987, pp. 1–17.

"Recent Progress in Comparative Neuroimmunology," *Zoological Science,* December 1992, pp. 1097–1100.

"What the Roaches Told Her," *The New York Times Magazine,* December 31, 1995, p. 26.

Florence Seibert

Born October 6, 1897
Easton, Pennsylvania
Died August 23, 1991
St. Petersburg, Florida

American biochemist Florence Seibert is best known for her research of tuberculosis, an infectious respiratory disease commonly known as TB. She developed the protein substance used for the tuberculosis skin test, which diagnoses the disease and which was adopted as the standard by the World Health Organization in 1942. In the early 1920s Seibert also discovered that the sudden fevers that sometimes occurred during intravenous injections (injecting blood, medicine, or other solutions directly into a person's vein) were caused by bacteria in the distilled water (water that had been purified through evaporation and condensation but was not sterile) used to make the protein solutions. To prevent contamination she invented a distillation apparatus that made distilled water almost completely sterile. This innovation was especially significant when intravenous blood transfusions became widely used in surgery. Having contracted polio as a child, Seibert overcame physical limitations to make important contributions to the field of biochemistry.

American biochemist Florence Seibert developed PPD-S, the protein substance used for the skin test to diagnose tuberculosis.

IMPACT

Florence Seibert developed PPD-S (Purified Protein Derivative-Standard), a tuberculin substance that was adopted in 1942 and is still in use today to diagnose tuberculosis. Since the introduction of the test, TB has been effectively controlled, at least in North America and Europe, with antibiotic drugs. But the disease is now spreading once again at an alarming rate. The World Health Organization has declared that it will claim more than thirty million lives by the year 2000 if immediate action is not taken. In many developing nations about three million people—nine thousand each day—are dying from TB as the result of new drug-resistant strains and the AIDS (acquired immunodeficiency syndrome) epidemic. AIDS attacks the body's immune system, leaving people more susceptible to diseases such as TB.

Studies chemistry

Florence Barbara Seibert was born on October 6, 1897, in Easton, Pennsylvania. She was the second of three children born to George Peter Seibert, a rug manufacturer and merchant, and Barbara Memmert Seibert. At age three Seibert contracted polio (poliomyelitis, an acute viral infection that causes paralysis and shrinking of skeletal muscles, often resulting in permanent disability and deformity). Despite the handicaps caused by the disease, Seibert completed high school with the help of her supportive parents. After graduation she attended Goucher College in Baltimore, Maryland, where she studied chemistry and zoology (a branch of biology concerned with the classification and study of animal life) and earned a bachelor's degree in 1918. Under the direction of one of her chemistry teachers, Jesse E. Minor, Seibert worked at the chemistry laboratory of the Hammersley Paper Mill in Garfield, New Jersey. Seibert and Minor had responded to a call for women to fill positions vacated by men fighting in World War I (1914–18). They also coauthored scientific papers on the chemistry of cellulose (a glucose substance that makes up the cell walls of plants) and wood pulps (chemically produced wood materials used in making paper).

Invents distilling apparatus

Although Seibert initially wanted to pursue a career in medicine, she was advised against it because it was thought that such an endeavor was "too rigorous" for a person with her physical disabilities. She therefore chose to study biochemistry and entered graduate school at Yale University in New

Haven, Connecticut. At Yale she studied under Lafayette B. Mendel, one of the discoverers of vitamin A. For her Ph.D. research she investigated the causes of "protein fevers," which often developed in patients after they had been injected with protein solutions that contained distilled water. (Proteins are essential to the structure and functioning of all living cells.)

Seibert's assignment was to discover which proteins caused the fevers and why. After extensive experimentation she found that the distilled water itself, not the proteins, was the source of the fevers, the reason being that the distilled water itself was contaminated with bacteria (microorganisms that cause disease). To prevent bacterial contamination she invented an apparatus for distilling water. Because Seibert's apparatus was able to collect and filter larger water droplets, less bacteria was able to survive the process, and as a result the distilled water was almost completely sterile.

Upon completing her doctorate in 1923 Seibert was awarded a fellowship at the University of Chicago. Under the direction of H. Gideon Wells she continued her work on pyrogenic (fever-causing) distilled water. Her work in this area had practical significance when intravenous blood transfusions (the transfer of blood into a vein of a person) became a standard part of many surgical procedures.

Begins tuberculosis research

After her fellowship ended, Seibert was employed part-time at the Otho S. A. Sprague Memorial Institute in Chicago, where Wells was the director. At the institute she also worked with Esmond R. Long, whom she had met through Wells's seminars at the university. Supported by a grant from the National Tuberculosis Association, Seibert and Long would eventually spend thirty years collaborating on tuberculosis research. Another of Seibert's long-time associates was her younger sister, Mabel, who moved to Chicago in 1927 to become her research assistant. For the rest of their lives, with the exception of a year Seibert spent in Sweden, the sisters resided together. In 1932 Seibert joined Long at the Henry

Robert Koch, German Bacteriologist

One of the founders of the field of bacteriology, Robert Koch (1843–1910) pioneered methods for studying bacteria and discovered the agents that cause tuberculosis, cholera (a gastrointestinal disease), and anthrax (a bacterial lung disease). Koch is also regarded as a founder of the public health movement, which promoted legislation and information about hygiene to prevent the spread of infectious diseases. In 1880 he was appointed an advisor to the Imperial Department of Health in Berlin. His task was to develop methods of isolating and cultivating disease-producing bacteria and to formulate strategies for preventing the spread of these bacteria.

The following year Koch published a report outlining four basic criteria, now known as Koch's postulates (assumptions), that are essential for identifying disease-causing bacteria. First, the organism or bacteria must be found in the tissue of animals with the disease, not in disease-free animals. Second, the organism must be isolated from the diseased animal and grown in a pure culture in vitro (outside the body). Third, the in vitro organism must be able to be transferred to a healthy animal, who will subsequently show signs of infection. And fourth, the organism must be able to be isolated from the infected animal. Six months later Koch succeeded in isolating a bacillus (disease-producing bacterium) from tissues of humans and animals infected with tuberculosis. When he presented his important findings before the Physiological Society in Berlin on March 24, 1881, he held the audience spellbound. That day has come to be known as the date of the birth of modern bacteriology. For his work on tuberculosis, Koch was awarded the Nobel Prize in 1905.

Phipps Institute, a tuberculosis clinic and research facility at the University of Pennsylvania in Philadelphia.

Seibert's tuberculosis research involved questions that had emerged from the work of German bacteriologist Robert Koch (see box). In 1892 Koch had discovered that the tubercle bacillus is the primary cause of tuberculosis, a fatal disease that had reached epidemic proportions in the late nineteenth century. In pulmonary tuberculosis, the most common form of the disease, infection attacks the respiratory system (the lungs and related nervous and blood supply systems) in two stages. First, small, hard, dormant (temporarily inactive) masses called tubercles are formed on the lungs by the body's defenses. Then the dormant bacteria in the tubercles are reactivated as a result of the weakening of the defenses, resulting in fatigue, weight loss, and a persistent coughing up of yellow sputum (saliva and mucus) and possibly blood.

Confirms Koch's hypothesis

Koch found that if the liquid on which the tubercle bacilli grew were injected under the skin, a small bite-like reaction, such as reddening or swelling around the injection site, would occur in people who had been infected with the disease. Calling the liquid "old tuberculin," he produced it by cooking a bacterium culture and draining off the dead bacilli. Although he believed the active substance in the liquid was protein, he had not been able to prove his hypothesis. Expanding on Koch's experiment, Seibert used precipitation (separation of a substance from a solution in the form of a solid) and other methods of separation and testing on the liquid. She discovered that the active ingredient was indeed protein. Her next task was to isolate the protein so it could be used in pure form as a skin test for tuberculosis. A skin test was necessary because tuberculosis is not easy to diagnose from symptoms, since the cough and fever brought on by the disease could easily be dismissed as cold symptoms. Proteins are highly complex organic molecules that are often difficult to purify, so Seibert had a challenging task. She finally succeeded in isolating the protein by using crystallization (changing the substance to crystal form).

The tiny amounts of crystal Seibert obtained, however, made them impractical for use in widespread skin tests. Thus she changed the direction of her research and began working on larger amounts of active, but less pure, protein. Her methods included precipitation through ultrafiltration (a method of filtering molecules). The result, after further purification procedures, was a dry powder called TPT (Tuberculin Protein Trichloroacetic acid precipitate). This was the first substance that could be produced in sufficient quantities for widespread use as a tuberculosis test. Continuing her experiments on tuberculin molecules, Seibert began working with the "old tuberculin" created by Koch and used by doctors for skin testing. Further purification of the substance led to the development of PPD (Purified Protein Derivative), and soon large quantities of PPD were being produced for tuberculosis testing.

Develops test for use throughout the world

In 1937 Seibert was awarded a one-year Guggenheim fellowship to study at the University of Upsala in Sweden. At Upsala she worked in the laboratory of Theodor Svedberg, who received the 1926 Nobel Prize for his protein research. While studying with Svedberg she learned new techniques for the separation and identification of proteins in solution, which aided her research on purifying and understanding the nature of protein. When she returned from Sweden in 1938, Seibert brought the new techniques with her. That year she also received the Trudeau Medal from the National Tuberculosis Association for her work on tuberculosis.

Upon returning to the Phipps Institute Seibert began work on producing a large batch of PPD that could be used as the basis for a standard dosage in skin testing. Establishing a standard dosage was critical because it was necessary to measure skin sensitivity to the test. In addition, it was important to determine whether sensitivity was the result of an individual skin reaction or differences in PPD substances. A large amount of PPD was also required for use throughout the world, so that the test would be the same wherever it was given. Devising new methods as the experiment proceeded, Seibert and her colleagues eventually produced 107 grams of material known as PPD-S (the S signified "standard"). A portion of the substance was used in 1941 as a U.S. government standard for purified tuberculins. The following year PPD-S was adopted as the standard tuberculosis skin test throughout the world.

The tuberculosis skin test is still the most common method used to diagnose the tuberculosis disease. During the skin test, the PPD-S is administered under the skin of a person's arm with a tine (prong) which has multiple needles. In two to three days if there is a red swelling at the site of the needle prick, the test is positive and indicates the presence of tuberculosis. This early detection of the disease aids in effective treatment of it.

Honored for achievements

During her twenty-seven-year career at the Henry Phipps Institute, Seibert advanced from the position of assistant professor of biochemistry in 1932 to the rank of full professor in 1955. After she retired in 1959 she and her sister moved to St. Petersburg, Florida. Seibert continued her research at a nearby hospital and in her home, concentrating on bacteria associated with certain types of cancers. She was the author or coauthor of more than one hundred scientific papers. In addition to the Trudeau Award, her numerous honors included five honorary degrees, the Garvin Medal of the American Cancer Society, and the John Eliot Memorial Award of the American Association of Blood Banks. Seibert's declining health in the last two years of her life was attributed to complications from her childhood polio. She died in St. Petersburg, Florida, on August 23, 1991.

Further Reading

Bailey, Brooke, *The Remarkable Lives of 100 Women Healers and Scientists,* Bob Adams, Inc., 1994.

New York Times, August 31, 1991.

Seibert, Florence, *Pebbles on the Hill of a Scientist* (self-published), 1968.

William Shockley

Born February 13, 1910
London, England
Died August 12, 1989
San Francisco, California

American physicist William Shockley was responsible for the invention of the transistor.

American physicist and inventor William Shockley led the team of scientists who, in 1947, invented one of the most important electrical devices of the modern age: the transistor. As early as 1939 Shockley was working with semiconductors in an attempt to improve the obsolete vacuum tube. The transistor, which was the result of Shockley's work, sparked the computer revolution and won Shockley the Nobel Prize in 1956. Despite this great accomplishment, however, Shockley's reputation was tarnished by his belief that intelligence is related to race, which drew intense criticism during the Civil Rights movement of the 1960s.

Studies solid-state physics

William Shockley was born on February 13, 1910, in London, England, to American parents. His father, William Hillman Shockley, worked in England as a mining engineer, and his mother, May Bradford Shockley, was a mineral surveyor. When Shockley was three the family returned to Palo Alto, California. Because his parents believed in home schooling, Shockley did not enter elementary school until the age of eight. His interest in physics developed early, inspired in part by a neighbor who taught the subject and by his parents' coaching and encouragement. By the time he had graduated from Palo Alto Military Academy and Hollywood High School at age seventeen, Shockley was committed to a career in physics. After attending the University of California at Los Angeles for one year, he earned his undergraduate degree in physics from the California Institute of Technology (Cal Tech) in Pasadena in 1932.

After graduation Shockley won a teaching fellowship at the Massachusetts Institute of Technology (MIT) in Cambridge, Massachusetts, where he also began his graduate work. His doctoral thesis examined the behavior of electrons in crystals, part of an area of study called "solid-state" physics (a technology that uses semiconductor devices rather than electron tubes). Shockley earned his doctoral degree in 1936.

Works on vacuum tubes

After turning down offers from General Electric and Yale University, Shockley took a job at the Bell Telephone Laboratories in Murray Hill, New Jersey. His first assignment was the development of a new type of vacuum tube (an electronic device in which all or most of the gas has been removed, permitting electrons to move with low interaction) that would serve as an amplifier (something that intensifies a weak current). But, almost as soon as he arrived at Bell, he began to think of a radically new approach to the transmission of electrical signals using solid-state components rather than conventional vacuum tubes. At that time vacuum tubes were used in most communication devices because they have the ability to rectify (change from alternating current to direct current) and multiply electronic signals. But they have a number of serious disadvantages, too. They are fragile and burn out quickly, making them expensive to operate on a large scale, and they are inefficient. Like lightbulbs, vacuum tubes heat up as an

> **IMPACT**
>
> When the transistor was first announced in a short article in the *New York Times* on July 1, 1948, few readers had the vaguest notion of the impact William Shockley's fingernail-sized device would have on the world. Before the transistor, computers used so many bulky vacuum tubes that several rooms were needed to house them. Today's microchips, based on transistor technology, are millions of times smaller, cheaper, longer-lasting, and more efficient and are found in products ranging from radios and televisions to weapons systems.

electric current passes through them; this factor signals a lack of efficiency, since heat is actually wasted energy.

Consequently, Shockley began thinking about other ways of handling electrical currents. By 1939 he was focusing on semiconductors, substances made of silicon and germanium that conduct an electrical current much less efficiently than do conductors such as silver and copper, but more effectively than do insulators such as glass and most kinds of plastic. Shockley knew that one semiconductor, galena, had been used as a rectifier in early radio sets, and his experience in solid-state physics led him to believe that such materials might have even wider application in new kinds of communication devices.

Shockley's research on semiconductors was interrupted by the outbreak of World War II. From 1940 to 1945 he became involved in military research, first developing radar equipment at a Bell field station in New Jersey and later as research director of the U.S. Navy's Anti-Submarine Warfare Operations Research Group at Columbia University. After the war he returned to Bell Labs as director of its research program on solid-state physics. Together with John Bardeen, a theoretical physicist, and Walter Brattain (see boxes), an experimental physicist, Shockley returned to his study of semiconductors.

Continues semiconductor research

Semiconductors are basically bits of silicon (a nonmetallic element) or related material, similar to a grain of sand in chemical makeup. Shockley believed that such material could be used as a switch, that its conducting ability could be controlled to change the flow and direction of an electrical current. He was also convinced that it could act as an amplifier,

Walter Brattain, American Physicist

In 1936 American physicist Walter Brattain (1902–1987) had been engaged in a variety of projects, such as studying infrared radiation and magnetic phenomena, while working at Bell Telephone Laboratories in New Jersey. With the arrival of William Shockley, however, his work took a new direction. Shockley had been attempting to find a substitute for the bulky three-vacuum-tube core that made up the modern radio. Shockley designed a number of arrangements containing semiconductors, substances whose ability to carry electrical current is lower than that of a conductor (like metal) and higher than that of insulators (like rubber), that he thought might work. Brattain's job was to build and test each of these arrangements.

For about ten years Brattain studied semiconductor surfaces. Finally, in 1947 he and John Bardeen, another physicist involved in the project, built a successful device. Brattain, Bardeen, and Shockley realized, however, that the new device, which they called a transistor, could be improved, and Shockley suggested a modified device that Brattain built and successfully tested by 1950. In 1956 Brattain, Bardeen, and Shockley shared the Nobel Prize for physics for their invention of the transistor.

allowing a weak current to be made much stronger. Until this time, switching and amplifying an electrical current had been accomplished only with vacuum tubes.

Invents the transistor

After more than a year of trial and error, Bardeen suggested that the movement of electric current was being hampered by electrons trapped within a semiconductor's surface layer. The team decided to suspend temporarily its efforts to build an amplification device in order to concentrate instead on learning all they could about the nature of semiconductors. By 1947 Bardeen and Brattain had learned enough about semiconductors to make another attempt at building Shockley's device. This time they were successful. Their device consisted of a piece of germanium with two gold contacts on one side

John Bardeen, American Physicist

American physicist John Bardeen (1908–1991) has the unique distinction of being the first—and as of the mid-1990s the only—person to have won two Nobel Prizes in physics. The first, which he shared with William Shockley and Walter Brattain, came in 1956 for his role in the discovery of the transistor. Only a few months after receiving the award, Bardeen completed another research project that would lead to his second Nobel in 1972. For this one he was recognized for his part in developing the BCS theory of superconductivity, named for the last names of its three inventors: Bardeen, Leon Cooper, and J. Robert Schrieffer, who shared the prize. The BCS theory accounts for the tendency of certain materials to lose all electrical resistance when they are cooled to temperatures close to absolute zero (the point at which matter stops moving; usually –273°C). The theory not only applied to all known superconductivity phenomena, where an electric current will persist indefinitely in a metal without any voltage, but also suggested an extensive agenda of new research on the topic. The BCS theory is often described as one of the most important developments in theoretical physics since quantum theory (the study of the various energy levels contained in atoms).

and a tungsten contact on the opposite side. When an electrical current was fed into one of the gold contacts, it appeared in a greatly amplified form on the other side. The device was given the name transistor (for *trans*fer re*sistor*). More specifically, it was referred to as a point contact transistor because of the three metal contacts used in it.

Receives Nobel Prize

Most of Shockley's discoveries were made in December 1947 and announced at a press conference in June of the following year. A few months later, after determining that the transistor's wire contact points could easily be broken or dislodged, Shockley proposed a modification. He suggested using a thin layer of P-type semiconductor (with a deficiency of electrons) sandwiched between two layers of N-type semicon-

William Shockley, Walter Brattain, and John Bardeen at the Bell Telephone Laboratories in 1948.

ductor (with a surplus of electrons). When Brattain built this device, now called the junction transistor, he found that it worked much better than did its point contact predecessor. This design took two years to perfect, and it was 1951 before it began appearing in consumer items such as radios, microphones, and hearing aids. The junction transistor quickly became the backbone of the electronics industry, and in 1956

Shockley, Brattain, and Bardeen won the Nobel Prize for physics for their invention.

Goes to Silicon Valley

In 1954 Shockley left Bell Labs to take advantage of the business opportunities his invention offered. He, his wife, and three children returned to California, settling in Palo Alto, where Shockley had lived as a child. There he set up Shockley Transistor Corporation. As the computer revolution gained momentum, other electronics companies followed Shockley to this area of California, giving it the nickname "Silicon Valley." After selling his company for a large profit in the early 1960s, Shockley became a professor at nearby Stanford University, where he taught electrical engineering until his retirement in 1975.

Stirs controversy

At Stanford Shockley became interested in genetics and the origins of human intelligence, in particular the relationship between race and IQ (intelligence quotient). Although he had no background in psychology, genetics, or any related field, he began to read on these topics and formulate his own hypotheses. He sparked considerable controversy when he suggested that intelligence is related to race. Shockley's theory centered on the idea that intelligence is passed on genetically and that whites naturally possess a higher degree of intelligence than nonwhites. The social implications of Shockley's theories were—and still are—deeply felt, particularly in the midst of the Civil Rights movement of the 1960s, when African Americans were staging an intense struggle for equality. Consequently, students at Stanford protested in his classes, his reputation in the scientific community became tarnished, and the media began to question his motivations.

Shockley had been awarded many honors during his lifetime, including the U.S. Medal of Merit in 1946, the Comstock Prize of the National Academy of Sciences in 1954, and the Institute of Electrical and Electronic Engineers Gold Medal in 1972 and its Medal of Honor in 1980. He was named to the National Inventor's Hall of Fame in 1974. Shockley died of prostate cancer on August 12, 1989, having cast a shadow on race relations while leaving a legacy of scientific advancement.

Further Reading

Aaseng, Nathan, *The Inventors: Nobel Prizes in Chemistry, Physics and Medicine,* Lerner, 1988.

Aaseng, Nathan, *Twentieth-Century Inventors,* Facts on File, 1991.

Maxine Singer

Born February 15, 1931
New York, New York

American biochemist Maxine Singer has been instrumental in establishing guidelines for DNA research.

American biochemist and geneticist Maxine Singer is a staunch advocate of responsible use of biochemical genetics research. During the height of the controversy over using deoxyribonucleic acid (DNA) techniques to create new genetic characteristics, she advocated a cautious approach. (DNA is the substance that carries the genetic information in living cells from generation to generation.) She helped develop guidelines that limit genetics research in order to protect the public from possible harm. Following the DNA controversy, Singer continued to contribute to the field of genetics, researching cures for cancer, hemophilia, and other diseases related to genetics.

Begins career

Maxine Singer was born on February 15, 1931, in New York City, to Hyman Frank and Henrietta (Perlowitz) Frank. She received a bachelor's degree from Swarthmore College in

Pennsylvania in 1952, and earned her Ph.D. in biochemistry (the study of chemical substances in living organisms) from Yale in 1957. From 1956 to 1958 she worked as a U.S. Public Health Service postdoctoral fellow at the National Institute for Arthritis, Metabolism and Digestive Diseases (NIAMD), a division of the National Institutes of Health (NIH), in Bethesda, Maryland. She then became a research chemist, working on enzymes and cellular biochemistry from 1958 to 1974. During this time she conducted DNA research on tumor-causing viruses as well as on ribonucleic acid (RNA), the nucleic acid found in cells that functions in the synthesis of proteins. In the early 1970s Singer also served as a visiting scientist with the Department of Genetics of the Weizman Institute of Science in Rehovot, Israel.

Scientists discover "recombinant DNA"

While Singer was working at NIH, scientists learned how to take DNA fragments from one organism and insert (recombine) them into the living cells of another. This "recombinant DNA" could direct the production of proteins in the second organism as if the DNA were still in its original home. This technique had the potential of creating completely new types of organisms. On one hand, the new research brought unprecedented opportunities to discover cures for serious diseases, to develop new crops, and to otherwise benefit humanity. Yet the prospect of creating as-yet-unknown life forms, some possibly hazardous, was frightening to many people, both scientists and nonscientists.

Warns public of gene-splicing risks

In 1972 one of Singer's colleagues and personal friends, **Paul Berg** (see entry) of Stanford University in California, was the first to create recombinant DNA molecules. He later voluntarily stopped conducting DNA experiments in the genes of tumor-causing viruses because of some scientists' fears that an unknown virus might escape from the laboratory and spread into the general population. Although Berg's action to stop con-

IMPACT

American biochemist and geneticist Maxine Singer provided a calm voice of reason during the public debate over the safety of "recombinant DNA" technology. An angry outcry arose when the potential results of DNA manipulation were compared in the media to the nightmarish vision in Aldous Huxley's novel *Brave New World,* which describes a genetically altered society. Singer advocated a careful analytic approach to the problem by arguing that freedom of inquiry is a democratic right, and that the new research could benefit humanity with the discovery of new cures for serious diseases. However, she also warned against irresponsible use of DNA technology, advocating guidelines to regulate laboratory research.

ducting his research was significant, the debate over gene-splicing was actually sparked at the 1973 Gordon Conference, an annual high-level research meeting. Singer, who was co-chairperson of the event, was approached by several nucleic acid scientists with the suggestion that the conference include consideration of safety issues. Singer agreed. She opened the discussion with an acknowledgment that DNA manipulation could assist in combating health problems, yet such experimentation raised a number of moral and ethical concerns.

The scientists at the conference decided, by ballot, to send a public letter about the safety risks of recombinant DNA research to the president of the National Academy of Sciences (NAS) and then publish the letter in *Science* magazine. Concern generated by the letter, which also suggested that the NAS recommend research guidelines, led to another meeting and a lengthy debate at the Asilomer Conference Center in Pacific Grove, California. Such a discussion of ethical issues arising from research was unprecedented in the scientific community. Immediately after the Asilomer Conference, a NIH committee began formulating guidelines for recombinant DNA research.

Creates controversy

While helping develop the guidelines in 1976, Singer advocated a careful approach to DNA research. She advised that certain experiments posed such serious hazards that they should be banned altogether. Singer was willing to permit experiments with lesser or no potential hazards if their benefits are unobtainable through conventional methods, and if they are properly safeguarded. She warned that the more risk involved in an experiment, the stricter the safeguards should be. She also recommended that the guidelines be reviewed annually.

Singer provided a voice of reason throughout the public debate that followed. Committees of nonscientific people, such as the Coalition for Responsible Genetic Research, held demonstrations calling for a complete ban on recombinant DNA research. Newspaper and television reporters made comparisons to Aldous Huxley's book *Brave New World,* which described a genetically altered society. When Singer addressed a public forum on the issue in 1977, she warned "scientists [to] recognize their responsibility to the public."

Stresses benefits of DNA technology

During her career Singer has also served on the editorial board of *Science* magazine and has contributed numerous articles. In writing for that publication she stressed the benefits to humanity that recombinant DNA techniques could bring, especially for increasing the understanding of serious and incurable diseases. After the NIH guidelines were implemented, she told *Science* readers that "under the Guidelines work has proceeded safely and research accomplishments have been spectacular." By 1980 the public near-hysteria had waned, and Singer called for a "celebration" of the progress in molecular genetics. She wrote in *Science* that "the manufacture of important biological agents like insulin and interferon by recombinant DNA procedures" as well as the failure of any "novel hazards" to emerge was evidence of the value of the cautious continuation of DNA research.

Appointed head of cancer research lab

In 1974 Singer accepted a new position at NIH as chief of the Section of Nucleic Acid Enzymology, Division of Cancer Biology and Diagnosis (DCBD) at the National Cancer Institute in Bethesda. In 1980 she became chief of the DCBD

Laboratory of Biochemistry. She held this post until 1988, when she became president of the Carnegie Institution, a research organization in Washington, D.C. Singer remains affiliated with the National Cancer Institute, however, as scientist emeritus, where she continues her research in human genetics.

Promotes the "Flavr Savr" tomato

In addition to her laboratory work, Singer has devoted considerable time and energy to other scientific and professional pursuits. In 1981 she taught in the biochemistry department at the University of California at Berkeley. In addition to her work with *Science* magazine, she has published more than one hundred books, articles, and papers. Singer also compiled a graduate-level textbook with Paul Berg on molecular genetics titled *Genes and Genomes: A Changing Perspective.* She and Berg also wrote a book for the general audience on genetic engineering. Singer continued to promote the benefits of recombinant DNA techniques and battle public suspicion and fear long after the controversy peaked in the 1970s. In the early 1990s, for example, she wrote an article encouraging the public to try the "Flavr Savr" tomato, the first genetically engineered food to reach American supermarket shelves. Responding to public objections that eating the tomato was dangerous, unnatural, or immoral, Singer said that the small amount of extra DNA would be destroyed in the digestive tract. She also pointed out that people already consume the DNA present in other foods in their diets.

Remains active in the scientific community

Singer has served on numerous advisory boards in the United States and abroad, including science institutes in Naples, Italy, Bangkok, Thailand, and Rehovot, Israel. She also has served as a consultant to the Committee on Human Values of the National Conference of Catholic Bishops. Concerned about the quality of science education in the United States, she started First Light, a science program for inner-city children. Singer is the recipient of more than forty honors and awards, including some ten honorary doctor of science degrees and numerous commendations from NIH. She married Daniel Singer in 1952, and they have four children.

Further Reading

Hoffee, Patricia A., "The New Genetics," *Science,* December 13, 1991.

Lappé, Marc, *Broken Code: The Exploitation of DNA,* Sierra Club Books, 1984, pp. 19–25.

"Making a Difference: The Year's Class Acts," *U.S. News & World Report,* August 26, 1991.

"The Scientific Method," *U.S. News & World Report,* August 26, 1991, p. 94.

Singer, Maxine, "Seeing a Red Menace in the New Hot Tomato," *Asbury Park Press,* August 15, 1993.

Singer, Maxine, and Paul Berg, *Dealing With Genes: The Language of Heredity,* University Science Books, 1992.

Singer, Maxine, and Paul Berg, *Genes and Genomes: A Changing Perspective,* University Science Books, 1990.

Soares, Marcelo Bento, "Precise Genetic Concepts," *Bioscience,* March 1992, p. 211.

B. F. Skinner

Born March 20, 1904
Susquehanna, Pennsylvania
Died August 18, 1990
Cambridge, Massachusetts

"The ideal of behaviorism is to … apply controls by changing the environment in such a way as to reinforce the kind of behavior that benefits everyone."

American behavioral psychologist B. F. Skinner was on the cutting edge of behaviorism, a group of theories that sets out to prove that human and animal behavior is based not on independent motivation but rather on response to reward and punishment. He entered the field of behavioral psychology, part of a larger branch of science known as social science, in the 1920s, when it was a relatively new area of study. Many scientists considered Skinner's lab experiments to be exciting and innovative, revealing an understanding of human behavior patterns. Yet these same experiments made him notorious among other colleagues who thought his views of human nature were too rigid and even inhuman. Skinner was famous for the invention of the Skinner Box, a scientific apparatus that was used to condition behavior.

Begins long academic career

Burrhus Frederic Skinner was born on March 20, 1904, in Susquehanna, Pennsylvania, the son of William Arthur

Skinner, an attorney, and Grace Burrhus Skinner. As a child he wanted to be a poet and novelist, but found he "had nothing to say." Skinner attended Hamilton College in Clinton, New York, earning a bachelor of arts degree in 1929. He went on to Harvard University in Cambridge, Massachusetts, where he received a master of arts degree in 1930 and a doctorate in 1931. After receiving his doctorate he served a one-year appointment as a research fellow with the National Research Council before returning to Harvard in 1933. Three years later Skinner became an instructor at the University of Minnesota in Minneapolis, rising to the rank of associate professor of psychology in 1939, and holding the position until 1945. Also in 1936 Skinner married Yvonne Blue. They had two daughters, Julie and Deborah. From 1942 to 1943 Skinner conducted war research for the U.S. Office of Scientific Research and Development.

In 1945 Skinner moved to the University of Indiana at Bloomington to become professor of psychology and department chairperson. He returned again to Harvard in 1948, remaining there until his retirement in 1974, when he was named professor emeritus. During the last sixteen years of his tenure at Harvard he was the Edgar Pierce Professor of Psychology.

Designs Skinner Box

Skinner derived his behaviorist beliefs from a series of laboratory experiments he conducted with rats and pigeons during the 1930s and 1940s. By rewarding his test animals whenever they performed a desired behavior (a process he called positive reinforcement), Skinner succeeded in training them to do a number of difficult tasks. His pigeons could play Ping Pong, dance, walk in figure eights, and distinguish colors. He taught rats to push buttons, pull strings, and press levers to receive food and drink. Skinner's research convinced him that behavior control could be achieved through the control of the environment. To provide a perfectly controlled environment in which to condition the animals, Skinner designed an enclosed, soundproof box equipped with tools, levers, and other devices. Widely used by other researchers, this training environment became known as the Skinner Box.

Designs Air-Crib

Convinced that the techniques he used on his pigeons and rats could work on human beings as well, Skinner built a training box for children in 1943. He called it the Air-Crib, and it was dubbed the "baby box" by the media. The glassed-in crib with carefully controlled temperature was designed, in Skinner's words, to provide a "very comfortable, stimulating environment" so that a baby need not wear constricting clothing. Skinner raised his own daughter, Deborah, in the box for two and a half years. However, when his account of the child-rearing experiment was published in the *Ladies' Home Journal,* it sparked a national controversy. Skinner was accused of monstrous experiments with his own children. Newspaper editors attacked him, and stories about him were featured on radio shows and in newsreels (short news reports on film). He replied to the critics by stating that his daughter spent no more time in the box than other infants did in their cribs or playpens. In fact, she suffered no ill effects from the experiment and grew up normally, remaining on good terms with her father until his death. Nonetheless, Skinner's attempts to market the "baby box" under the name "Heir Conditioner" was a failure.

Writes a handbook for the 1960s

Skinner used his laboratory work and experiments with his daughter's upbringing to formulate a theory for dealing with social problems. He concluded that the same techniques that successfully train laboratory animals can be used to control negative human behavior and eliminate such social ills as crime, poverty, and war. In 1948 Skinner applied his theories in *Walden Two,* a novel in which he portrays a society based on behaviorist principles, with positive and negative reinforcements built into the social structure. The plot revolves around a tour of the hypothetical community taken by two college professors. They observe that children, raised in communal nurseries, are taught to think and learn instead of being called upon to memorize specific facts. The teaching of religion and history is suppressed, and all members of the society encourage social harmony by practicing positive reinforcement for approved behavior.

Walden Two was welcomed by many social scientists (people who study human society and individual relationships in society) as an insightful model for creating a better society, but critics warned of the potential for totalitarianism (total control of a society by a few people in positions of power) in Skinner's ideas. They called him a "social engineer" or, less politely, a "neo-fascist." (A fascist is a person who believes in a system of government marked by dictatorship, government control of money, and suppression of all opposition.) About twenty years later, during a period of widespread social change in the United States, *Walden Two* met with a more sympathetic and receptive audience. It became a handbook for the "flower children" of the 1960s, young people who rejected materialism and advocated a society based on peace, love, and community. They were attracted by Skinner's vision of a utopia (perfect society) in which communes (a small group of people with similar interests and who often share property) of people have destroyed evil and replaced it with rewarding work and unlimited artistic self-expression. The book was translated into eight languages and sold a million copies.

IMPACT

American behavioral psychologist B. F. Skinner developed the theory that humans are controlled (stimulated) solely by forces in their environment. According to behaviorist concepts, rewarded behavior (positive reinforcement) is encouraged, and unrewarded behavior (negative reinforcement) is terminated. The ideas and findings of Skinner and his behaviorists have had a tremendous impact on such areas as drug and alcohol rehabilitation, where the chief concern is behavior modification. Skinner stands out from other behaviorists in that he not only dismissed the scientific analysis of the human thinking process, but he also believed that "feelings and mental processes are just the meaningless byproducts of [the] endless cycle of stimulus and response." Skinner's ideas about behavior influenced the design of such simple items as crib toys for babies and such complex processes as inventory management systems for industry.

Proposes technology of behavior

In 1971 Skinner wrote *Beyond Freedom and Dignity,* which has been described as the major achievement of his career. Whereas he had given a fictional account of his ideas in *Walden Two,* in *Beyond Freedom and Dignity* he openly argues for social change based on behaviorist theory. Skinner observes that "almost all our major problems involve human behavior, and they cannot be solved by physical and biological technology alone. What is needed is a technology of behavior." A technology of behavior would utilize knowledge about the interaction between the individual and the environment to design a society capable of changing destructive human behavior through a system of positive and negative reinforcements. But before this can come about, Skinner cautioned that we must reject the belief that humans are self-directed and can act freely.

Reactions to *Beyond Freedom and Dignity* were divided. Social scientists and critics who were opposed to behaviorist theory thought Skinner's denial of human autonomy (independence) was misguided. Echoing earlier reactions to the Skinner Box and the Air-Crib, many considered his plans for a controlled society to be dangerous. Skinner's supporters, however, disputed that his views were not totalitarian, and that in fact he proposed a society in which people themselves have power over social control mechanisms.

Writes autobiography

Skinner also addressed social problems in his three-volume autobiography—*Particulars of My Life* (1976), *The Shaping of a Behaviorist* (1979), and *A Matter of Consequences* (1983). Each book covers a particular period of his life. The first volume traces his childhood and education, the second describes his research work during the 1930s and 1940s, and the third gives an account of his later life as one of the leading psychologists of the time. Skinner's autobiography contributed to the debate—which continues even today—about applying experiments done on laboratory animals to humans in order to change behavior and structure a new modern society.

Although Skinner's eyesight was limited near the end of his life, he wrote *Upon Further Reflection* in 1987 and *Recent*

Issues in the Analysis of Behavior, which was published in 1989. Skinner died in Cambridge, Massachusetts, on August 18, 1990. During his career he received numerous honors and awards, including the National Medal of Science (1968), the American Psychological Association gold medal (1971), and the Humanist of the Year Award from the American Humanist Society (1972). Skinner was also presented honorary degrees from twenty-five colleges and universities.

Further Reading

Bjork, Daniel, *B. F. Skinner: A Life,* Basic Books, 1993.

Chicago Tribune, October 14, 1990.

People, August 13, 1990; October 29, 1990.

Skinner, B. F., *Beyond Freedom and Dignity,* Knopf, 1971.

Skinner, B. F., *Walden Two,* Macmillan, 1948; published with introduction by Skinner, Macmillan, 1976.

Washington Post, October 21, 1990.

Patrick Steptoe

Born June 9, 1913
Oxfordshire, England

Died March 21, 1988
Canterbury, England

English gynecologist and medical researcher Patrick Steptoe helped develop the process of in vitro fertilization.

English gynecologist Patrick Steptoe was best known for helping develop the technique of in vitro fertilization, a process by which an egg is fertilized by sperm outside of the body. Steptoe and his colleague, English physiologist Robert G. Edwards, received international recognition—both positive and negative—when the first so-called test tube baby was born in 1978.

Begins practicing medicine

Patrick Christopher Steptoe was born on June 9, 1913, in Oxfordshire, England. His father was a church organist, and his mother was a social worker. Steptoe studied medicine at St. George Hospital Medical School at the University of London and, after being licensed as a physician in 1939, became a member of the Royal College of Surgeons. His medical career was interrupted by World War II (1939–45), when he volunteered as a naval surgeon. During a sea battle in 1941, Steptoe and his shipmates were captured by Italian forces after their ship sank. Although he was at first granted special privileges in prison because he was a physician, he was placed in solitary confinement when his captors caught him helping other prisoners escape. Steptoe was released as part of a prisoner exchange agreement in 1943. Following the war he completed additional studies in obstetrics (a branch of medicine that deals with birth) and gynecology (a branch of medicine concerned with disorders of the female reproductive system). In 1948 he became a member of the Royal College of Obstetricians and Gynecologists and moved to Manchester to set up a private practice. In 1951 Steptoe began working at Oldham General and District Hospital in northeast England.

Uses laparoscope

While at Oldham Steptoe pursued his interest in fertility problems, or the inability of a woman and man to conceive a child. He developed a method of removing human eggs from the female's ovaries (the female reproductive glands that produce eggs) by using a laparoscope, a long, thin microscope with a fiber optics light. After inserting the device through a small incision made in the navel and into the inflated abdominal cavity, Steptoe was able to observe the female reproductive tract. Eventually, the laparoscope would also become widely used in various types of surgery, including those associated with sterility (the inability to produce offspring). However, at first Steptoe had trouble convincing others in the medical profession of the merits of laparoscopy; observers from the Royal College of Obstetricians and Gynecologists decided the technique was too problematic. As a result, five years passed before Steptoe published his first paper on laparoscopic surgery.

Pioneers in vitro fertilization

In 1966 Steptoe teamed up with Cambridge University physiologist Robert G. Edwards to advance his work with fertility problems. (A physiologist is a scientist who studies organs, tissues, and cells.) Utilizing ovaries removed for med-

IMPACT

English gynecologist Patrick Steptoe developed in vitro fertilization, a method of obtaining mature human eggs from female ovaries, fertilizing them outside the body, then replanting them in a woman's uterus. As a result of Steptoe's pioneering techniques, the first "test tube baby" was born on July 25, 1978. Since then couples with various physiological problems have had children in fertility clinics throughout the world.

ical reasons, Edwards had pioneered the fertilization of eggs outside of the body, called in vitro fertilization (IVF). The process of in vitro fertilization is controversial as well as inventive. In this operation, a mature egg (an egg ready to be fertilized) is removed from the female ovary and is fertilized with sperm in a test tube. After a short incubation or development period, the fertilized egg is implanted in the uterus, where it develops as in a typical pregnancy. (The uterus is the organ in the female reproductive system where a fertilized egg develops.) This procedure gives hope to women who cannot become pregnant because their fallopian tubes (a pair of tubes that guides eggs from the ovaries to the uterus) are damaged or missing.

With his laparoscope Steptoe added the dimension of being able to secure mature eggs at the appropriate moment in a woman's monthly menstrual cycle when fertilization would normally occur. A breakthrough for the scientists came in 1968, when Edwards successfully fertilized an egg that Steptoe had extracted. Not until 1970, however, were they able to produce an egg that had reached the stage of cell division—into about one hundred cells—when it generally moves and implants in the uterus during a typical pregnancy. In 1972 Steptoe and Edwards attempted to implant an embryo (the earliest stage in an organism's development) in a uterus, but the embryo failed to lodge in the uterus. In fact, none of the women with implanted embryos carried them for a full trimester (a period of three months in the nine-month pregnancy cycle).

Steptoe and Edwards encounter opposition

As news of their work began to spread, Steptoe and Edwards faced intense criticism. Scientists and religious leaders raised ethical and moral questions about tampering with

"Baby Louise" Brown, First Test Tube Baby

Joy Louise Brown, the world's first test tube baby, came to be known as "Baby Louise" by millions of newspaper readers and television viewers throughout the world. Stories about her birth brought both fame and notoriety to Patrick Steptoe, who perfected in vitro fertilization. Brown's family lived under constant scrutiny by the media for several years. Today Brown lives in Bristol, England, with her parents and her sister Natalie, who was also a test tube baby. In 1994 she told a *People* magazine interviewer that she rarely thinks about the unusual circumstances surrounding her conception, and that she plans to attend college to become a pediatric nurse.

the creation of human life. Many critics compared their efforts to the strictly controlled society depicted by Aldous Huxley in his novel *Brave New World* (1932). In Huxley's futuristic scenario babies are conceived in a laboratory, cloned, and manipulated for society's use. Members of British Parliament demanded an investigation, and the scientists' research funds were cut off. Nevertheless, Steptoe and Edwards continued their work at Kershaw's Cottage Hospital in Oldham, with Steptoe financing the research by performing legal abortions, the termination of a pregnancy. In order to avoid criticism, they became more secretive, but this secretiveness only managed to make the speculation and criticism even more intense.

Produce first test tube baby

In 1976 Steptoe met Leslie Brown, a thirty-year-old woman who had experienced problems with her fallopian tubes and wanted to have a baby. Steptoe removed a mature egg from her ovary, and Edwards fertilized the egg in a petri dish using the sperm of her husband, Gilbert. After two days, Steptoe implanted the fertilized egg in Brown's uterus, where it continued to thrive as a normal pregnancy. On July 25, 1978, Brown gave birth to a healthy 5-pound 12-ounce baby girl at Oldham hospital. The Browns named the baby Joy Louise (see box).

Cause controversy and outrage

Reluctant to discuss the IVF birth publicly, Steptoe and Edwards did not immediately publish their findings in a medical journal. In October 1978 Steptoe was to receive an award from the Barren Foundation, a fertility research organization based in Chicago, Illinois. However, the foundation suddenly canceled the presentation because Steptoe and Edwards had not published an article about their procedure, although by this time stories about "Baby Louise, the First Test-Tube Baby" were appearing in newspapers and on television throughout the world. In addition, rumors were rampant that Steptoe and Edwards had sold their story to the tabloid *The National Enquirer* for a large sum of money. Steptoe declared he had rejected such offers and did not make any money on the highly publicized birth.

Steptoe reports on in vitro procedure

Defending his decision not to publish an article in a scientific journal immediately after the test tube birth, Steptoe answered that most scientists do not publish until several months after data are in and research is complete. He subsequently presented the full details of the procedure at the conference of the American Fertility Society in San Francisco, California. He reported that with modified techniques, 10 percent of the IVF attempts could succeed, and predicted that one day there could be a 50 percent success rate for the procedure. Despite the furor Steptoe and Edwards had generated, the New York Fertility Society presented Steptoe an achievement award.

Receives awards and honors

In the aftermath of the first successful test tube baby, Steptoe received thousands of letters from couples seeking help in conceiving a child. He retired from the British National Health Service and constructed a new clinic near Cambridge. For their efforts, Steptoe and Edwards were both named commanders of the British Empire, and in 1987 Steptoe was hon-

In Vitro Fertilization in the 1990s

Since 1978, when Patrick Steptoe's pioneering in vitro fertilization (IVF) techniques resulted in the birth of the first test tube baby, it has become commonplace throughout the world for couples to have children with the aid of fertility clinics. Obstacles that prevent these couples from having children include advanced age, men's poor sperm quality, and the rejection of healthy embryos (the earliest stage of development) by women's bodies. Although reports show that the failure rate for fertility treatments is higher than the success rate, scientists are working on new technologies to improve the treatments.

At Monash University in Melbourne, Australia, for example, Alan Trounson has helped develop an IVF method called immature oocycle collection. Unlike the standard IVF procedure developed by Steptoe and Robert G. Edwards, in which only mature eggs are removed from a woman's ovaries to be fertilized in the lab, immature oocycle collection allows doctors to remove immature eggs and bring them to maturity outside the ovary. Trounson predicts his method will reduce IVF cost by 80 percent.

Another development is intracytoplasmic sperm injection (ICSI), which enables men with low sperm counts or abnormal sperm to father children. Perfected by embryology professor Gianpiero Palermo, the technique involves injecting a single sperm into the egg. Since 1992 more than one thousand babies have been born by means of ICSI, and Palermo reports a 38 percent success rate. Although the cost is twice as high as that of standard IVF, scientists predict ICSI will soon be used in all IVF procedures.

IVF continues, however, to raise questions about the ethics of controlling the birth process and "designing" an embryo—an even more disturbing possibility in the minds of some critics. For instance, a team of geneticists headed by Eugene Pergament at Prentice Children's Hospital in Chicago, Illinois, has been working on a way to determine the sex of embryos used for IVF. But most controversial is preimplantation genetic diagnosis (PGD), a tool developed at George Washington University in Washington, D.C. By analyzing the deoxyribonucleic acid (DNA) of an embryo before implanting it in the uterus, the test can detect possible genetic disorders. More than thirty women with high genetic risks have given birth to healthy babies after PGD in the United States and Europe. Critics are concerned, however, that the test could be used to produce embryos with specific genetic characteristics. Thus the ethical issues raised by Steptoe's test tube baby are still troubling scientists and the general public today.

ored with a fellowship in the Royal Society. Steptoe and his wife, a former actress, had a son and a daughter. Steptoe died of cancer on March 21, 1988, in Canterbury, England.

Further Reading

Beaubien, Greg, "Progress Against Infertility," *American Health,* September 1995.

Cowley, Geoffrey, "The Future of Birth," *Newsweek,* September 4, 1995.

"From Miracle Baby to Regular Teen," *People Weekly,* February 7, 1994.

Gabriel, Trip, "High-Tech Pregnancies Test Hope's Limit," *New York Times,* January 7, 1996.

Gibbs, W. Wayt, "Fertile Ground," *Scientific American,* February 1994.

Gillis, Anna Maria, "Finding the Right Sperm for the Job," *BioScience,* September 1995.

Grady, Denise, "Unnatural Selection," *Vogue,* October 1995.

Steptoe, Patrick, and Robert Edwards, *A Matter of Life: The Story of a Medical Breakthrough,* Morrow, 1980.

Thompson, Larry, "Fertility With Less Fuss," *Time,* November 14, 1994.

Nikola Tesla

Born July 10, 1856
Smiljan, Croatia
Died January 7, 1943
New York, New York

Serbian-American inventor and electrical engineer Nikola Tesla was the first person to prove and perfect the efficient use of alternating-current electricity, a regular flow of electrons that moves first in one direction and then another. His polyphase system, which produced a steady flow of electricity, become the standard for power transmission throughout the world. He also pioneered research in such areas as artificial lightning, high-frequency currents, high-tension currents, and radio telegraphy. Before his death in 1943 Tesla had acquired more than one hundred patents. His famous Tesla coil and other inventions have since become integral to modern technology.

Nikola Tesla was the first scientist to perfect the use of alternating-current electricity.

Nikola Tesla, the son of Serbian parents, was born on July 10, 1856, in the Croatian village of Smiljan, in an area that later became Yugoslavia. Tesla's father and mother, Milutin Tesla and Djuka Mandic, had expected him to follow in his father's footsteps as a Greek Orthodox clergyman. But Tesla excelled in math and science during his early school years in Smiljan and then in the nearby town of Gospic, where his parents moved when he was six or seven. Gradually it became clear that the young and independent-minded Tesla was not a candidate for the seminary.

Challenges the Gramme dynamo

In 1871, when Tesla was fifteen, he attended the higher secondary school at Karlovac, Croatia. After four years he moved to Graz, Austria, to enroll at the polytechnic institute there, and again excelled in math and science. He had an exceptional memory (he was reported to have memorized *Faust,* the drama by Johann Wolfgang von Goethe) and showed particular interest in electrical engineering. One day, while attending a class where a Gramme dynamo (a direct-current induction motor) was being demonstrated, Tesla noticed unnecessary and potentially dangerous sparks being released from the dynamo; he also noticed that the higher the voltage that was being generated, the greater was the arcing of the sparks. Tesla commented that these sparks were caused by the motor brushes coming into contact with the commutator (a switching device that causes a current to reverse direction). He suggested the sparks could be eliminated by creating a motor that did not have a commutator. The professor was skeptical of the young scientist's theory, so nothing came of the idea at the time. Over the coming years, however, Tesla would continue to work to overcome the problems of motors using direct current, or a regular flow of electrons always moving in the same direction.

Begins working in Edison plants

Tesla received his university education in Prague, Czechoslovakia, but the details of his life there are unclear. However, his post-Prague years come into sharper focus. In January 1881 Tesla moved to Budapest, where he worked in the Hungarian government's new central telegraph office. During his brief tenure there Tesla invented a telephone amplifier (which intensifies a weak electrical current), yet for reasons unknown he never patented the device. He also continued to think about the sparks created by the Gramme dynamo in the classroom in Graz. During this time he studied rotating magnetic fields, which would later become the basis for all polyphase induction motors. The following year Tesla took a position with the Continental Edison Company in Paris, France. His job was to correct problems encountered at Edison plants in Germany and France.

In 1884 Tesla decided to move to the United States, where there were interesting developments in electrical engineering and more opportunities to receive funding for his research. He secured a position at the Edison research laboratory in New York City. Although **Thomas Alva Edison** (see entry) had already made a reputation for himself as an electronics wizard, he was committed to the use of direct-current electricity. Tesla began a difficult relationship with his stubborn boss when Tesla explained to Edison his plans for a motor based on alternating current, which he found to be less problematic than direct current and capable of higher voltage. Edison insisted that Tesla's designs for this new motor were impractical and dangerous. Nevertheless, Edison hired Tesla, and for a year the new immigrant designed direct-current dynamos and motors for the Edison Machine Works in New York. Tesla was unhappy with his job, however, finding he was unable to overcome the personal and professional differences with Edison.

IMPACT

Although Nikola Tesla's experimentation with high energy and high voltage electrical currents produced no immediate practical results, his studies provided the basis for research by later scientists. Physicist Robert Golka, for example, modeled his work in plasma physics on material he gleaned from Tesla's notes, which were housed at the Tesla Museum in Belgrade after World War II. Similarly, Soviet physicist Pyotr Kapitsa, who shared the 1978 Nobel Prize for his research on magnetism, acknowledged Tesla's work as a model for his own research. Richard Dickinson, a researcher at the California Technical Institute Jet Propulsion Laboratory who experimented with transmission of wireless energy, also used Tesla's concepts as a guide.

Founds Tesla Electric Company

The following year some entrepreneurs persuaded Tesla that he could start his own electric company in Rahway, New Jersey. Tesla saw an opportunity to work out his ideas for alternating current in a practical way, but his financial supporters seemed mainly interested in providing arc lighting—a stream of brilliant light produced when electric current jumps across to electrodes surrounded by gas—for streets and factories. Once again he faced disappointment and was forced to work for at least part of 1886 as a common laborer. Nevertheless Tesla continued to work on his innovations in his spare time, and during this period he managed to acquire seven patents for arc lamps. Tesla's growing interest in electrical innovations gradually worked to his advantage, and by 1887 he was able to establish the Tesla Electric Company.

Creates first efficient polyphase motor

As a new company owner Tesla invented the first efficient polyphase motor. His design incorporated several wire-taped blocks that surrounded the rotor (the rotating part of the motor). When alternating current is supplied to the wires, the current to each block is slightly out of phase with the others, and a rotating magnetic field is created. The movement of the rotor is achieved as it follows this revolving field. The practical effect of Tesla's invention was that it allowed strong electrical currents to be transmitted over long distances. Edison's direct current motor, on the other hand, was limited to local use and required many electrical relay stations to distribute the current throughout an area such as a city.

Disproving Edison's assertion that alternating current was impractical, Tesla would acquire forty patents related to this technology by 1891. As his inventions began to attract attention in the late 1880s, he began giving lectures. Perhaps the most notable was a presentation he gave to the American Institute of Electrical Engineers in May 1888 that earned him a reputation as a brilliant electrical engineer.

Begins working for Westinghouse

Manufacturer and inventor George Westinghouse bought one of Tesla's patents for the polyphase motor, then hired Tesla to work at his plant in Pittsburgh, Pennsylvania. In 1889 Tesla became an American citizen. He was now famous and his future seemed assured. During the ensuing years he contin-

Nikola Tesla photographed circa 1895 with one of his electrical generators.

ued his research and lectured to prestigious organizations throughout the United States and Europe. In Britain he addressed the Institution of Electrical Engineers and the Royal Society, and in France, the Society of Electrical Engineers and the French Society of Physics. In these lectures Tesla discussed his work in the transmission of electrical power through radio waves (electromagnetic waves with radiation energy). At the Columbian Exposition in Chicago in 1893, Westinghouse provided electricity with Tesla's system of polyphase alternating current, making it the first world's fair to use electricity. Tesla also gave lectures and demonstrations of his research at the exposition.

Designs world's first hydroelectric generating plant

Tesla's partnership with Westinghouse gave him the opportunity to design what may have been the scientist's greatest achievement: the world's first hydroelectric generating plant. Located in Niagara Falls, New York, the plant distributes electrical current to Niagara Falls and to Buffalo, New York, some twenty-three miles away. The Niagara power plant, completed in 1895, proved the workability of Tesla's polyphase system of alternating current and established the kind of power system that would eventually be used throughout the United States and the world.

Turns attention to radio

In 1897 Tesla turned his attention to the idea that radio waves could carry electrical energy, a process that involves transmitting electricity without wires. He successfully achieved wireless communication over a distance of twenty-five miles. The following year he demonstrated the transmission of electrical energy with several radio-controlled model boats he had constructed. Since the Spanish-American War was underway at the time, there was little public recognition of his new revelation, and possibly this type of remote-control system was too advanced for its usefulness to be fully appreciated. It would also take some time before some of Tesla's other inventions would be found to be beneficial. His work with high-frequency currents yielded several generating machines that were forerunners of those used in radio communication. His Tesla coil, a resonant (vibrating) air-core transformer, proved capable of producing currents at higher frequencies and magnitudes.

Creates artificial lightning

In 1898 Tesla moved to Colorado Springs, Colorado, where he continued his experiments on electricity, but this time on a larger scale than model boats. As before, his interests focused on transmission of high energy, sending and receiving wireless messages, and related issues pertaining to high voltage electricity. He built a 200-kilowatt transmitting tower that could produce lightning bolts so powerful they could overload the city's electrical generator. In fact, during one experiment in creating artificial lightning, Tesla caused the municipal generator to catch fire, plunging the town into darkness.

Almost wins Nobel Prize

Late in 1915 newspapers printed rumors that the Nobel Prize committee was considering Tesla and Edison as candidates to share the Nobel Prize in physics. Tesla became enraged that he might have to share the prize with his arch rival. For reasons never made clear, however, the committee gave the award in physics to two other candidates. In 1917 one of Tesla's colleagues recommended him for the prestigious Edison Medal of the American Institute of Electrical Engineers. Again, because of the association with Edison, Tesla refused the award, then finally agreed to accept it. When he attended the award banquet held in his honor, he soon drifted from the crowd and was later found outside feeding the pigeons.

Dies in New York City

During the last four decades of his life Tesla became reclusive, living alone in a room at the Hotel New Yorker in New York City. His influence within the scientific community soon faded, and his colleagues worried that he was beginning to lose his grasp for rigorous scientific inquiry. For example, Tesla announced that while he was at Colorado Springs he had received radio signals from intelligent life on Mars or Venus. Although radio signals from space are now a normal part of astronomical research, in the early years of the twentieth century such an idea was considered ludicrous. Furthermore, on his seventy-eighth birthday Tesla told an interviewer he had plans for an invincible death beam that could instantly destroy ten thousand airplanes or one million soldiers. He publicly offered to create such a death beam for the U.S. government in three months for less than $2 million.

On January 8, 1943, a maid discovered Tesla's body in his hotel room. He had been ill for two years and had evidently died in his sleep of a coronary thrombosis, a disease in which a clot forms in the heart and destroys the muscle. He was eighty-six years old. After his death Tesla finally received the recognition that had not come to him during his life. Scores of notable people—President Franklin Roosevelt and his wife Eleanor, New York mayor Fiorello H. La Guardia, political figures from Yugoslavia, Nobel Prize winners, leaders in science—praised him as a visionary who provided the foundations for modern technology. In fact, within a year of his death, the United States Supreme Court ruled that Tesla, and not **Guglielmo Marconi** (see entry), had invented the radio. Yugoslavia made him a national hero and established the Tesla Museum in Belgrade after World War II. In addition to honorary degrees from American and foreign universities (including Columbia and Yale in 1894), and the Edison Medal, during his life Tesla received the John Scott Medal. In 1975 he was inducted into the National Inventors Hall of Fame.

Further Reading

Cheney, Margaret, *Tesla: Man Out of Time,* Dorset Press, 1981.

Dictionary of Scientific Biography, Volume 13, Scribner, 1976, pp. 286–87.

Dommermuth-Costa, Carol, *Nikola Tesla: a Spark of Genius,* Lerner Publications, 1994.

Hall, Stephen S., "Tesla: A Scientific Saint, Wizard or Carnival Sideman?," *Smithsonian,* June 19, 1986, pp. 121–34.

Lawren, Bill, "Rediscovering Tesla," *Omni,* March 1988, pp. 65–66, 68, 116–17.

Neidle, Cecyle S., *Great Immigrants,* New York, 1973.

Tesla, Nikola, *The Inventions, Researches and Writings of Nikola Tesla,* originally published in *The Electrical Engineer,* 1894, reprinted by Barnes & Noble, 1992.

Vivien Thomas

Born in 1910
Nashville, Tennessee
Died in 1985

Vivien Thomas was highly regarded in the medical community for his scientific genius and surgical skill. With no formal medical training, he helped develop intricate surgical techniques that ultimately saved thousands of lives. Thomas performed research in the animal laboratories at Vanderbilt University during the 1930s, leading to the widespread use of blood and plasma transfusions during World War II. Later, at Johns Hopkins University, he worked with heart surgeon Alfred Blalock to perform hundreds of experimental procedures on laboratory dogs to develop the "blue baby" operation for treating congenital cyanotic heart disease. Although he received little formal recognition during Blalock's lifetime, Thomas served as supervisor of the surgical research laboratories at Johns Hopkins from 1941 until 1979. During this time he played an important role in training many of the nation's top surgeons.

"To have an honorary degree conferred upon me was far beyond any hope or expectation I could imagine."

IMPACT

Despite his never fulfilling his dream of becoming a medical doctor, Vivien Thomas had far-reaching influence on the medical community. As supervisor of the surgical research laboratories at Johns Hopkins University for nearly four decades, Thomas developed surgical techniques used by his long-time associate, prominent heart surgeon Alfred Blalock, as well as by the many top surgeons that he trained. Thomas conducted thousands of experimental procedures in the laboratory, a few hundred alone in developing the surgical technique that would be used to save thousands of "blue babies," infants with congenital cyanotic heart disease, which gave the skin of infants a bluish cast. For his contributions to medicine Johns Hopkins awarded Thomas an honorary doctorate in 1976; the following year he was made an official member of the medical school faculty.

Dreams of medical career

Vivien T. Thomas was born the son of a building contractor in Nashville, Tennessee, in 1910. During his childhood he was impressed by the wisdom and kindness of his family doctor and made up his mind to pursue a career in medicine. After school, on weekends, and during the summer he worked as an apprentice carpenter and as an orderly in a private infirmary, hoping to earn enough money to pay his college tuition. Thomas graduated from high school in 1929, and the following fall enrolled as a premedical student at Tennessee Agricultural and Industrial State College. Within a few months, however, the stock market had crashed, causing the banks to fail as well. Nearly all of Thomas's savings were gone. His dreams of an education were shattered.

Becomes lab assistant

The Great Depression also meant the end of carpentry projects in Nashville, so Thomas was forced to look elsewhere for permanent work. After working for several months as a jack-of-all-trades, repairing roofs, fixing plumbing, and mending worn-out steps, he heard about an opening in the research laboratory at Vanderbilt University Medical School. A surgeon named Alfred Blalock needed an assistant. Blalock was conducting groundbreaking research into the causes and effects of shock (the body's reaction to trauma). Because his clinical duties required him to spend much of his time with patients in the hospital, he wanted someone knowledgeable and independent who could carry out his research work in the laboratory. Although Thomas was only nineteen, Blalock was impressed by his confident, businesslike air and hired him immediately. Their association was to last until Blalock's death in 1964.

Thomas had originally intended to work in the Vanderbilt laboratory part-time, and only until he had earned enough money to return to Tennessee State. After completing his studies, he hoped to attend Meharry Medical College. But as the Depression deepened, his prospects for further education grew dimmer. "For the time being, I felt secure," he wrote in his autobiography. "At least I had a job. It seemed to be a matter of survival." He devoted himself to his work. Before long he was spending sixteen hours a day in the laboratory, supervising experiments at night and devouring chemistry and physiology textbooks during the day.

Conducts important research

Thomas's surgical talent manifested itself almost immediately as he assisted Blalock in experiments focused on hypertension, or high blood pressure, and traumatic shock. They worked to recreate in laboratory animals the kinds of stressful conditions people experienced when trapped for prolonged periods under piles of rubble. To help in the investigation Thomas invented a heavy spring device that they used to apply varying degrees of pressure to the extremities of animals. He then recorded each reaction. The information they collected enabled Blalock to formulate a new understanding of traumatic shock.

Blalock's landmark discovery that shock was linked to a loss of fluid and a decrease in blood volume earned him the respect and admiration of the international medical community and resulted in the widespread application of life-saving blood transfusions during World War II. As time went on Blalock spent less and less time in the laboratory, leaving Thomas with the responsibility for developing, monitoring, and calculating the results of all of their experimental procedures. Blalock trusted Thomas and regarded him as a full and equal partner. "It was extremely difficult to tell if Dr. Blalock had the original idea for a particular technique or if it was Vivien Thomas," a former intern at Johns Hopkins once remarked. "They worked so smoothly together, we—the medical students—didn't know."

Breaks racial barriers

In 1940 Blalock, then at the height of his career, was offered the position of surgeon in chief at Johns Hopkins Hospital. He accepted the job, but only on one condition: that Vivien Thomas be allowed to accompany him. Although the university agreed to Blalock's terms, Thomas's arrival in June 1941 caused quite a stir. According to an article in *Reader's Digest,* "On one occasion when Thomas walked across campus, he halted traffic. A black man in a lab coat was something unheard of at Johns Hopkins, which had segregated restrooms and a separate entrance for black patients. But inside the lab it was Thomas's skill, rather than his skin color, that raised eyebrows."

As director of the Hunterian Surgical Research Laboratory, Thomas continued to carry out research programs for Blalock. But it was Thomas's own inventive genius that helped bring many of the projects to fruition. Shortly after Blalock and Thomas arrived at Johns Hopkins, the hospital's chief of pediatrics approached Blalock for help in correcting the problem of coarctation (narrowing) of the aorta (the main artery carrying blood from the heart) in children. In order to investigate the disorder—as well as many others—in laboratory animals, Thomas devised a positive-pressure respirator that inflated the lungs of dogs under anesthesia. At that time, no such machine was available. He and Blalock then went on to create, and attempt to correct, specific cardiovascular (heart and blood vessel) defects in dogs.

Helps perfect surgical technique

In 1943 Johns Hopkins cardiologist Helen Taussig (see box) met with Blalock and Thomas to solicit their advice con-

Helen Taussig, American Physician and Cardiologist

American pediatrician and cardiologist Helen Taussig (1898–1986) realized that the blueness of cyanotic children, or "blue babies," was the result of insufficient oxygen in the blood. She had discovered that the condition was caused by either a leaking septum (the wall that separates the chambers of the heart) or an overly narrow artery leading from the heart's left ventricle to the lungs. Although at that time surgeons were unable to enter the heart to repair the septum surgically, Taussig believed it might be possible either to repair the artery or to attach a new vessel that would perform the same function.

Taussig approached Alfred Blalock, who had done experimental research on an artificial artery with his long-time assistant Vivien Thomas. Accepting Taussig's challenge, Blalock set Thomas to work on the technical problems. During the next year and a half Thomas developed the surgical procedures, using about two hundred dogs as experimental animals. In 1944, earlier than Thomas had planned, the technique was tried on one of Taussig's patients, a human infant. With Taussig as an observer and Thomas standing by to give advice concerning the correct suturing of the artery, Blalock performed the surgery successfully. A branch of the aorta that normally went to the infant's arm was connected to the lungs. In the years that followed, the procedure, known as the Blalock-Taussig shunt, saved the lives of thousands of cyanotic children.

cerning the treatment of an infant, Eileen Saxon, who had been born with deficient blood flow to the lungs. She was suffering from congenital cyanotic heart disease (a malformation of the heart that results in insufficient oxygen in the blood and causes a bluish discoloration of the skin and mucous membranes). She was only one of a growing number of "blue babies" whom Taussig had been unable to help. By this time Blalock and Thomas had performed hundreds of intricate cardiac procedures in the lab. In a series of experimental operations at Vanderbilt, they had attempted to bring more blood to the lungs of laboratory dogs by dividing a major artery and sewing it to the pulmonary artery that leads to the lungs. They believed that a refined version of this procedure might help Taussig's "blue babies."

Guides historic operation

Blalock and Thomas spent much of the following year developing and performing experimental procedures in the laboratory in preparation for the historic operation on Eileen Saxon. But it was Thomas who worked out the final details of the surgical technique and taught them to his famous associate. On the day of the operation, however, Thomas was nowhere to be found. Flanked by a surgical team that included some of the country's foremost physicians and researchers, Blalock refused to begin the procedure without his laboratory assistant at his side. Eventually Thomas was located in the laboratory and summoned to the operating floor. For the next three hours Thomas stood quietly at Blalock's right shoulder, watching carefully as the surgeon's scalpel and needles moved in and out and offering helpful suggestions. The operation was a success.

Because of Thomas's surgical expertise and his exhaustive knowledge of the procedure, Blalock insisted that he be present for the first one hundred operations. In fact, according to *Reader's Digest,* if any operating room staff member attempted to move into the space behind Blalock's right shoulder, the surgeon would deliver the quick admonishment, "Only Vivien is to stand there."

Trains physicians and technicians

In addition to providing invaluable assistance to Blalock in the operating room, as head of the surgical research laboratory at Johns Hopkins Thomas helped train a generation of surgeons and lab technicians. He taught them the latest techniques in vascular (vein) and cardiac (heart) surgery. He was also regarded as the Johns Hopkins veterinarian because of his success in treating both lab animals and pets belonging to students and staff. At one point, Thomas's reputation was such that practicing veterinarians in the Baltimore area called on him for consultation. Yet, despite his hard work and talent, Thomas remained largely invisible at Johns Hopkins, earning a low income. Only when he announced his intention to leave the university to accept a more lucrative job as a carpenter did the hospital offer to double his salary. Thomas then agreed to stay.

Honored by colleagues

By the time Blalock died in 1964 Thomas had come to terms with the fact that he had never obtained a college degree. He continued to supervise the surgical research laboratories at Johns Hopkins until his retirement in 1979. As time went on, however, increased teaching and administrative responsibilities prevented him from doing surgical research.

Thomas was finally honored for his work in 1971. He was genuinely surprised, and deeply moved, when a group of surgeons from around the country—many of whom he and Blalock had helped train—gathered in the auditorium at Johns Hopkins to pay tribute to his life and accomplishments. After the ceremony Thomas's portrait, which had been commissioned by his former students, was unveiled and hung next to a portrait of Blalock in the foyer of the Blalock Building. In 1976 he was awarded an honorary doctorate from Johns Hopkins, and the following year he became an official member of the medical school faculty. Shortly before his death in 1985, Thomas completed his autobiography, *Pioneering Research in Surgical Shock and Cardiovascular Surgery: Vivien Thomas and His Work With Alfred Blalock,* in which he recalled his life and work.

Further Reading

Harvey, A. McGehee, *Adventures in Medical Research,* Johns Hopkins University Press, 1976.

Journal of the American Medical Association, January 2, 1987, pp. 87–88.

Reader's Digest, October 1989.

Thomas, Vivien, *Pioneering Research in Surgical Shock and Cardiovascular Surgery: Vivien Thomas and His Work With Alfred Blalock* (autobiography), University of Pennsylvania Press, 1985.

William Thomson, Lord Kelvin

Born in 1824
Belfast, Ireland
Died in 1907

William Thomson, Lord Kelvin, is recognized as one of the great scientific thinkers of the nineteenth century. The originator of new schools of thought in physics, electronics, and mathematics, he also invented numerous scientific devices. Thomson was knighted in 1866 for his prodigious work, and in 1892 was made first baron Kelvin of Largs. It is by this name that his most influential discovery, the Kelvin scale of absolute temperature, is known.

William Thomson developed the theory of absolute temperature and established a temperature scale now named after him.

Enters college at age ten

Thomson was born in Belfast, Ireland, in 1824. The son of a respected mathematician, he was recognized as a child prodigy. When his father obtained a professorship at the University of Glasgow in Scotland, Thomson (then only eight years old) attended his father's lectures. At age ten he entered the University of Glasgow, as did his twelve-year-old brother James, studying mathematics and finishing second in his class.

IMPACT

William Thomson, Lord Kelvin, is remembered for inventing the Kelvin temperature scale, which defines thermodynamic temperature, or the level of heat in a gas, liquid, or solid. The absolute scale was easier for scientists to use than the centigrade scale since it took into account an absolute low temperature. The Scottish mathematician and physicist **James Clerk Maxwell** (see entry) utilized the absolute scale in his formulation of kinetic gas theory. In 1851 Thomson himself applied the scale to show that all energy is eventually converted into usable heat, and that heat is released into the atmosphere and lost. This line of thought influenced the concept of entropy (the degradation of the matter and energy in the universe to an ultimate state of inertia) formulated by German physicist Rudolf Clausius in his second law of thermodynamics in the 1850s.

While Thomson was still in his teens he enrolled in postgraduate courses, first at Cambridge University in England and then at the University of Paris in France. In 1846, at the age of twenty-two, he was appointed to the newly created position of professor of natural philosophy (an early term for science) at the University of Glasgow. He was responsible for the construction of Britain's first physics laboratory at the university.

Announces "true" age of Earth

The British scientific community eagerly awaited Thomson's first scientific publication as a professor; when he was in his early teens he had written a paper that was presented by a senior professor to the Royal Society of Edinburgh. Less than a year after his appointment Thomson announced his findings regarding the age of Earth, a number drastically lower than the calculation accepted by geologists at the time. He explained that by beginning with the assumption that the Sun and Earth were once the same temperature, he determined the number of years it would have taken for the planet to cool to its present temperature. The figure he arrived at was somewhere between 20 million years and 400 million years, most likely about 100 million years.

Previously geologists had decided that Earth was several billion years old, with the temperature remaining near normal for most of that time. These findings were eventually found to be much closer to the truth. Thomson seemed so convincing, however, that many scientists scrambled to fit their theories within Earth's new short life-span. Biologists in particular were dismayed because it meant that all of the evolution of life had to now fit within the span of one million years. Thomson's estimate also put forth the concept of evolutionary mutation. Until that time biologists believed that evolution followed a slow, steady path; now they were forced to consider that it involved "jumps" through which the process was accelerated. Although Thomson's theories were eventually disproved, the concept of evolutionary mutation remains.

Codevelops theory later known as Joule-Thomson effect

During his study of the age of Earth, Thomson became intrigued by the relationship between heat and energy. In 1847 he met English physicist **James Prescott Joule** (see entry), the author of some of the most innovative heat theories of all time. At that time, however, Joule's work was unknown, particularly in England. Thomson introduced Joule's theories to the Royal Society, a prestigious British scientific organization, gaining Joule the recognition he deserved. During the next few years Thomson and Joule worked together on experiments with the heat and energy of certain gases. One phenomenon they observed was that as a gas was introduced into a vacuum its temperature would drop and, if that drop were substantial enough, the gas would be converted to a liquid. Called the Joule-Thomson effect, this phenomenon became the basis for the liquification of most gases and, much later, the science of cryogenics (the branch of physics that involves the production and effects of very low temperatures).

Discovers absolute temperature

Thomson's work with Joule continued to stimulate his curiosity about the nature of heat. He was especially interested in the work of French physicist Jacques Charles, who had found that, for every degree centigrade (symbol C) below zero that a gas was cooled, the volume of the gas would decrease by 1/273. The implication of this theory was that at -273°C the volume of the gas would be zero. Scientists were unable to explain exactly why this would happen, just as they were unable to prove Charles's law to be false. In 1848 Thomson

Rudolf Clausius, German Physicist

William Thomson, Lord Kelvin, introduced the idea that all energy runs down and is dispersed into the environment as heat. This idea was later reintroduced by German physicist Rudolf Clausius (1822–1888) as the second law of thermodynamics, better known as entropy. In 1850, just three years after receiving his doctorate, Clausius published his influential essay on the degradation of energy. He proposed that this phenomenon could be expressed as the ratio of heat content to the absolute temperature. As the heat content within a system rose, so did the ratio, provided that the system was closed, thus allowing no energy to enter or escape. In such a system heat could be created but never removed, causing the ratio to increase until the entire system reached a static state.

Since scientists could not create a closed system in the laboratory, Clausius's concept remained purely theoretical. Nevertheless, his explanation did serve as proof of the inevitable degradation of energy. Fifteen years later Clausius labeled this concept entropy, a word derived from the Greek for "transformation," and it is now known as the Second Law of Thermodynamics, second in importance only to the knowledge that heat cannot move from a cool substance to a hot substance. In 1850, however, it was not universally welcomed by the scientific community. Only with the support of Scottish scientist James Clerk Maxwell were Clausius's theories eventually accepted.

explained the law: when the temperature of a gas is reduced, so is the energy level of the atoms in the gas; as the atoms move less they take up less room, thus decreasing the volume of the gas. At –273°C the energy level of the atoms reaches zero: the atoms stop moving, taking up almost no space, and their temperature cannot be lowered any further. Because this theory could be true for any substance, Thomson determined that –273°C is the absolute zero of temperature.

Invents absolute scale

In 1848 Thomson applied his theory by inventing an absolute scale of temperature. The scale essentially drops the centigrade scale by 273 degrees, so that zero and absolute zero coincide. Thus the Kelvin scale uses as its standard reference level absolute zero, rather than the freezing and boiling points of water, which are the standard references in both the metric (centigrade or Celsius) and the English (Fahrenheit) scales. In the metric system the difference between freezing and boiling is divided into one hundred equal intervals (degrees centigrade or Celsius). In the English system the intervals are divided into 180 units (degrees Fahrenheit). Thus the Kelvin (symbol K) has the same magnitude as the degree centigrade—the difference between freezing and boiling water in both scales is one hundred degrees—but the two temperature scales differ by 273.15 degrees. The freezing point of water on the centigrade scale is 0° and the boiling point is 100°. The freezing point of water on the Kelvin scale is 273° and the boiling point is 373°. Thomson called his scale the absolute scale, but after his death it was renamed the Kelvin scale.

Saves the transatlantic cable

During the mid-1800s the British scientific community was working on the first transatlantic telegraph cable. Thomson lent his knowledge of electrical theory to this effort. Maintaining that only very low voltages could transmit the telegraph signals at a sufficient rate over such a long cable, he invented a number of ultra-sensitive galvanometers, or instruments for detecting or measuring a small electrical current. Thomson's ideas clashed with the views of E. O. W. Whitehouse, an electrician who had been placed in charge of the project. Using a system of his own design, Whitehouse completed construction of a high-voltage telegraph cable. When Whitehouse's cable was tested in 1856, it proved to be a complete failure. After several years of litigation (lawsuits), the transmitters were replaced with Thomson's low-voltage system. Beginning in 1865 the underwater cable provided instant communication across the Atlantic. Recognized as the man who had rescued a gigantic financial investment, Thomson was knighted for his work in 1866. He was named Baron Kelvin of Largs in 1892.

Rejects theories of radioactivity

As the nineteenth century drew to a close Thomson lectured on the virtues of Victorian science, a school of thought that held that all the important discoveries in physics had been completed. So committed was he to his ideas that he rejected the theories of radioactivity, claiming no energy could be derived from a decaying atom. Despite his stubbornness Thomson was considered one of Britain's greatest scientists. He died in 1907 and was buried next to the great English mathematician and physicist Isaac Newton in Westminster Abbey in London, England.

Further Reading

Burchfield, Joe D., *Lord Kelvin and the Age of the Earth,* University of Chicago Press, 1990.

Sharlin, Harold, and Tiby Sharlin, *Lord Kelvin: The Dynamic Victorian,* Pennsylvania State University Press, 1979.

Smith, C. W., and M. N. Wise, *Energy and Empire: A Biographical Study of Lord Kelvin,* Cambridge University Press, 1989.

Thompson, Silvanus P., *The Life of Lord Kelvin,* 2 volumes, 2nd ed., Chelsea, 1977.

Tunbridge, Paul, *Lord Kelvin: His Influence on Electrical Measurements and Units,* Institution of Electrical Engineers, 1992.

Charles Townshend

Born April 18, 1674
Norfolk, England
Died June 21, 1738
Norfolk, England

Charles Townshend was a prominent British politician and wealthy landowner who developed innovative farming methods that quickly spread across eastern England. The ideas he promoted, including crop rotation and the cultivation of turnips to feed livestock, are as useful today as they were in the eighteenth century.

Charles Townshend's innovative farming methods have had a great impact on farm productivity since the Industrial Revolution.

Born on April 18, 1674, Charles Townshend was the oldest of three sons of Horatio Townshend, a viscount (nobleman) who had served in the British Parliament. The elder Townshend was one of the richest men in the eastern English county of Norfolk, and Charles, as oldest son, was heir to his father's substantial estate, Raynham. Accordingly, Townshend was provided with the best education possible. When he was eleven his parents sent him to Eton, a prestigious school for boys preparing for university study. A year later, upon the death of his father, Townshend became the new viscount.

In 1691 Townshend entered King's College at Cambridge University. During his college years he attended class sporadically and did not pursue a course of study leading to a degree or graduation. Instead he is said to have spent those years biding his time until he was old enough to take over the responsibilities of his father's title. At Cambridge he met Robert Walpole, the future prime minister of Britain. Then, after leaving college in 1694, Townshend headed for continental Europe, traveling for three years in Holland (the Netherlands), France, and Italy.

IMPACT

Charles Townshend's use of lease agreements to promote agricultural improvements was unusual, but highly effective, among eighteenth-century landowners. Townshend heightened the value of his property through demanding that his tenant farmers employ such methods as crop rotation and turnip cultivation. Although slowly at first, the new practices eventually caught on in Norfolk and neighboring counties, and Townshend's idea of planting turnips to feed livestock in winter improved England's economy.

Enters Parliament

Along with the title of viscount came a seat in the House of Lords. Upon returning from mainland Europe in 1697, Townshend assumed his duties in Parliament, at first taking the side of the conservative minority group, the Tories. However, he found that the positions of the Whig party were more to his liking, so he quickly changed parties. (The Whigs, who were then the ruling party in Britain, favored change and reform.) Townshend attended sessions of the House of Lords regularly and served on various committees.

In 1701 Townshend was made lord lieutenant, or chief military officer, of the county of Norfolk. This position allowed him to reward friends and supporters with commissions in the county militia. Townshend remained active in politics and often campaigned for Whig candidates seeking government offices. He later helped his brother Roger win a seat in Parliament.

Marries twice

While serving in the House of Lords, Townshend married his first wife, Elizabeth Pelham, daughter of Baron Thomas Pelham. In thirteen years of marriage the couple had several children; Charles, the eldest, would become the third Viscount

Townshend of Raynham. Elizabeth died in 1712, and Townshend did not remarry until he was fifty-two years old. His second wife, Dorothy Walpole, was the sister of his old school friend Robert.

Becomes diplomat to Holland

All of Townshend's political efforts were rewarded in 1709, when he was sent to Holland as a diplomat to negotiate a treaty with the Dutch. Since 1701 England had been fighting alongside Holland and several other countries against France and Spain in the War of Spanish Succession. While in Holland, Townshend successfully negotiated the "Barrier Treaty," by which England would support Holland's demand for a series of fortified towns in the Spanish Netherlands (now Belgium). Although the terms were later changed, the Barrier Treaty helped bring the war to a close.

Serves as secretary of state

In 1710 the Tories won control of Parliament in a landslide election. The Tory victory cost Townshend his diplomatic post in Holland as well as his title, lord lieutenant. He was not out of power for long, however. When George, the prince of the German state of Hanover, became King George I of England in 1714, Townshend was appointed secretary of state for northern affairs, the highest cabinet post at the time. He was responsible for foreign relations with the nations of northern Europe as well as some domestic affairs. His position also brought him within the king's inner circle of advisers.

The always outspoken and opinionated Townshend was soon quarreling with the king over foreign policy. George I maintained close ties to Hanover and wanted England to aid the state in its war with Sweden. Townshend opposed military intervention by Britain in the Baltic region. The dispute cost him the job of secretary of state in 1716, but he was appointed lord lieutenant (the king's chief representative) of Ireland. But five years later Townshend was again named secretary of state, a post he held until 1730. During the nine years of his second

Noel Davis, American Agronomist and Inventor

American agronomist and inventor Noel Davis has been experimenting since the 1970s with innovative agricultural methods that may be adopted by farmers of the future. On PhytoFarm, a one-acre farm inside an old, windowless warehouse near Dekalb, Illinois, Davis is growing spinach and herbs without the benefit of soil. In this controlled environment he uses electricity, water, conveyors, and computers to grow food that could someday be the new American breadbasket that feeds a hungry world. Known as hydroponics, this method relies on water instead of soil to grow plants. Water consumption is more economical at PhytoFarm, which uses five pounds of water to grow one pound of lettuce, compared with ten times that amount in an open, outdoor field. Davis also controls the temperature, humidity, light, carbon dioxide, and nutrients with a computer. A machine plants seeds tightly together in a polyester bed. When they need more space, the seedlings are transferred to conveyor belts. The plants inch along the conveyors and are fed a meal of nitrogen, phosphorous, potassium, and some trace elements, such as iron and calcium.

PhytoFarm has had mixed success since its inception in 1978. The cost of vegetables and herbs—under the brand name Kitchen Harvest—is still out of reach for most consumers. But nations with little farmable land such as Japan, huge populations like China, or cheap hydroelectric power as in Norway and Sweden are interested in Davis's indoor farm. Outer space may soon be another growing environment for PhytoFarm. Working with the Center for Space Automation and Robotics at the University of Wisconsin, Davis is studying how to grow fresh foods in low gravity to feed workers in outer space. The problem, so far, is finding ways for plant roots to stay in contact with the nutrient solution, which floats away because of the lack of gravity.

term he promoted a policy of neutrality in English disputes in the Baltic area and tried to recruit Austria as an ally against France. He also accompanied the king when he made occasional visits to his other court in Hanover.

Turns to farming

Meanwhile, Robert Walpole had become prime minister of Britain. In the later years of Townshend's term the two bickered increasingly over English policy. At the same time, Townshend's health was failing. In May 1730 he resigned as secretary of state and lord lieutenant of Ireland and left London to retire to his estate at Raynham. It was during his convalescence that he made remarkable contributions to the field of agriculture.

In eighteenth-century England most farming was done on small, unconnected strips of land in open fields—a practice dating back to feudal times. Traditionally, farmers planted a wheat crop, followed it with a barley crop, and then left the field fallow (unplanted) for a year in order to allow the soil to regenerate. Improved farming techniques were gradually introduced in the country. As early as the 1640s Sir Richard Weston, a refugee from the English civil war living in the Spanish Netherlands, noted that farmers improved their harvest by rotating their crops. Instead of letting the fields lie fallow between crops of grain and flax, the farmers planted turnips or clover; this not only replenished the soil but also provided feed for livestock.

Turnips and clover were introduced into England in the seventeenth century, and soon farmers such as Thomas Coke of Norfolk were experimenting with crop rotation. Jethro Tull, who lived in Townshend's time, used a seed drill to plant turnips instead of scattering the seeds over the whole field. The drill made it possible to plant crops in rows, thereby allowing for cultivation with horse-drawn hoes. Turnips grew well in the sandy soil of Norfolk, and Townshend had great success growing them on his estate. Thus he earned the nickname "Turnip Townshend."

Improves his estate

Although Townshend inherited a large estate from his father, much of it consisted of rush-covered marshes or sandy wastes that supported a few sheep. One historian described it as a place where "two rabbits struggled for every blade of grass." But Townshend took a keen interest in his land and was determined to enhance its quality and value. In the years before his departure as diplomat to Holland, he prepared the land for cultivation by creating drainage ditches, planting hedges, and fertilizing the soil with marl (a mixture of sand, clay, limestone, and seashell fragments). He is said to have eventually brought 400,000 acres under cultivation.

Promotes crop rotation

Townshend became an enthusiastic promoter of crop rotation after his retirement from politics. In lease agreements with tenants who worked portions of his land, Townshend demanded rotation: plant rye one year; leave the land fallow for a year; then rotate wheat, barley, and oats undersown with clover over the next three years. He also urged tenants to insert a crop of turnips among their grain crops during the rotation.

Through his ingenious methods Townshend greatly improved the value of his property. Crop rotation, turnip cultivation, and other agricultural improvements were gradually adopted by Norfolk farmers, though the new methods were slow to spread into surrounding counties. Gradually, however, new agricultural techniques made headway, and by the early nineteenth century farmers using Townshend's methods turned nearby Lincolnshire from rabbit patches and swampland into grain fields and pasture land. The introduction of turnips as livestock feed also improved England's pastoral economy by enabling farmers to feed their sheep during winter.

Townshend spent his last years at his mansion at Raynham with his two unmarried daughters. He died suddenly on June 21, 1738.

Further Reading

Rosenheim, James M., *The Townshends of Raynham,* Wesleyan University Press, 1989.

Russell, Sir E. John, *A History of Agricultural Science in Great Britain: 1620–1954,* Allen & Unwin, 1966.

James Van Allen

Born September 7, 1914
Mount Pleasant, Iowa

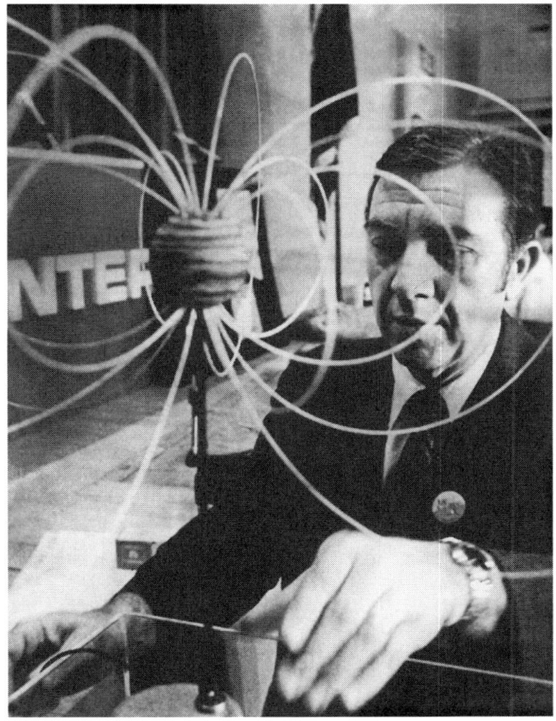

American physicist James Van Allen is best known for his discovery of high-level radiation bands that surround Earth. Popularly known as the Van Allen belts, these bands are part of the magnetosphere, a region dominated by Earth's magnetic field. Although Van Allen's discovery has been the highlight of his career, he has been associated with other significant scientific research. During World War II he created the radio proximity fuse, a small tracking device that triggers weapons when they are close to a target. Van Allen later applied his knowledge of rockets and satellites to the study of Earth's atmosphere.

American physicist James Van Allen played a central role in determining the composition of Earth's atmosphere.

Builds generator as a child

James Alfred Van Allen was born in Mount Pleasant, Iowa, on September 7, 1914. He was the second of four sons born to Alfred M. Van Allen, an attorney, and Alma E. (Olney) Van Allen. Interested in science from an early age, when he

IMPACT

James Van Allen is best known for discovering two bands of radiation, now known as the Van Allen belts, that encircle Earth. Part of the magnetosphere, the belts consist of charged particles that circulate along Earth's magnetic lines of force. These lines extend from above the equator to the North Pole, down to the South Pole, then back to the equator. The Van Allen belts, now taken into account when planning flights into outer space, provided a new basis for investigation of the upper atmosphere and contributed to the understanding of the aurora borealis.

was twelve he and his brother built their own electrostatic generator (a machine that produces high electrical voltages by accumulating large quantities of static electrical charges), with which they produced bolts of artificial lightning. By the time Van Allen reached high school, physics had become his passion. After graduating Van Allen entered Iowa Wesleyan College in his hometown. After receiving a bachelor of science degree in 1935, Van Allen entered the State University of Iowa at Iowa City for graduate study, where he earned a master of science degree in 1936 and a doctorate in 1939.

Conducts war research

When Van Allen's graduate work came to the attention of the Carnegie Institute in Pittsburgh, Pennsylvania, he was offered a position as research fellow in the Department of Terrestrial Magnetism. (Terrestrial means "of the earth"; magnetism is the property of a body to produce an electrical current around itself.) Spending three years at Carnegie, in 1942 he moved on to the Applied Physics Laboratory at Johns Hopkins University in Baltimore, Maryland. That same year, however, World War II became the dominant factor in Van Allen's life, and he left Johns Hopkins to accept a commission in the U.S. Navy, where he served until 1946.

Van Allen's first important scientific accomplishment, the radio proximity fuse, came as a result of his war research. The fuse was a device consisting of a radio transmitter and receiver that was attached to weapons. Signals sent out and received by the fuse indicated when the weapon was close to the target, allowing it to explode before actual impact. This greatly increased the efficiency of the missiles, as it eliminated the need for a direct hit.

Heads V-2 rocket program

Van Allen's work on the radio proximity fuse shaped his future scientific career in unexpected ways. At the end of World War II American scientists inherited about one hundred V-2 rockets built by German scientists. Far more advanced than any missiles produced by Americans at the time, these rockets promised to be valuable in the study of Earth's atmosphere. As a result of his research on the proximity fuse, Van Allen had become a leading authority on the miniaturization (design and construction in small size) of instruments, a skill crucial in rocket research. The U.S. Army therefore appointed him to administer and coordinate rocket-based research, and in 1946 Van Allen took charge of the V-2 research program based at the White Sands Proving Ground in New Mexico. His primary responsibilities were to design payloads to be sent into the atmosphere and to select projects using rocket research. (Payload is items considered not necessary to the purpose of a flight, such as scientific instruments or passengers.) From 1947 to 1958 he also served as chairperson of the committee overseeing this research, originally called the V-2 Rocket Panel and, later, the Rocket and Satellite Research Panel.

This change of name reflected a change in actual rocket research itself. With a limited number of V-2 rockets available, scientists soon began to explore alternative instruments. Although American scientists could not produce an equivalent of the V-2, under Van Allen's direction they eventually constructed an acceptable substitute, the Aerobee. The Aerobee carried 150 pounds of payload and reached altitudes of about 300,000 feet (about 60 miles). Over the next decade American scientists used the Aerobee in such projects as investigating and measuring solar radiation, sky brightness, atmospheric composition, and the aurora borealis (brilliant displays of bands or streamers of light visible in the night sky, chiefly in Earth's polar regions).

Returns to Iowa

In 1951 Van Allen returned to Iowa as professor of physics as well as the head of the department of physics and astronomy at the State University of Iowa, holding these positions until 1985. Shortly after returning to Iowa Van Allen developed a new approach for the study of the atmosphere. For over a century balloons had been the most effective way to accumulate information about the atmosphere. Instruments were carried into the atmosphere in balloons to heights of up to 15 miles and then parachuted back to Earth with their data. Van Allen combined ballooning techniques with modern rocket technology to make a "rockoon," which consisted of a balloon carrying a rocket. When the balloon reached its maximum altitude, the rocket was fired off by remote control. It traveled straight upward, through the balloon itself, another 50 to 70 miles into the atmosphere. Readings obtained from such rockoon launchings provided better information about the outer atmosphere than ever before.

Creates a new satellite

Information from two rockoons sent up in 1953 produced data that puzzled Van Allen. He discovered that levels of radiation at an altitude of 30 miles were much higher than had been expected. These results made him curious about what he might find at even higher altitudes. At this time scientists were discussing the next stage in rocket development: an artificial Earth satellite that would remain in long-term orbit around the planet. As part of the International Geophysical Year (a program that gathered together scientists from sixty-seven nations to study a variety of subjects) planned for 1957–58, the U.S. government made a commitment to finance such a satellite.

The first satellite in that program, *Explorer I,* was launched on January 31, 1958, carrying a Geiger counter (an electronic instrument that detects and measures radiation) that Van Allen had installed for measuring cosmic rays. Returned data showed that the counter registered increasing concentrations of cosmic radiation until it went above 600 miles, at which point no radiation was registered at all. The mystery continued until the experiment was repeated on *Explorer III,* this time with a lead shield to protect the Geiger counter from the majority of rays. The shield was a test of Van Allen's hypothesis that high levels of radiation were jamming the

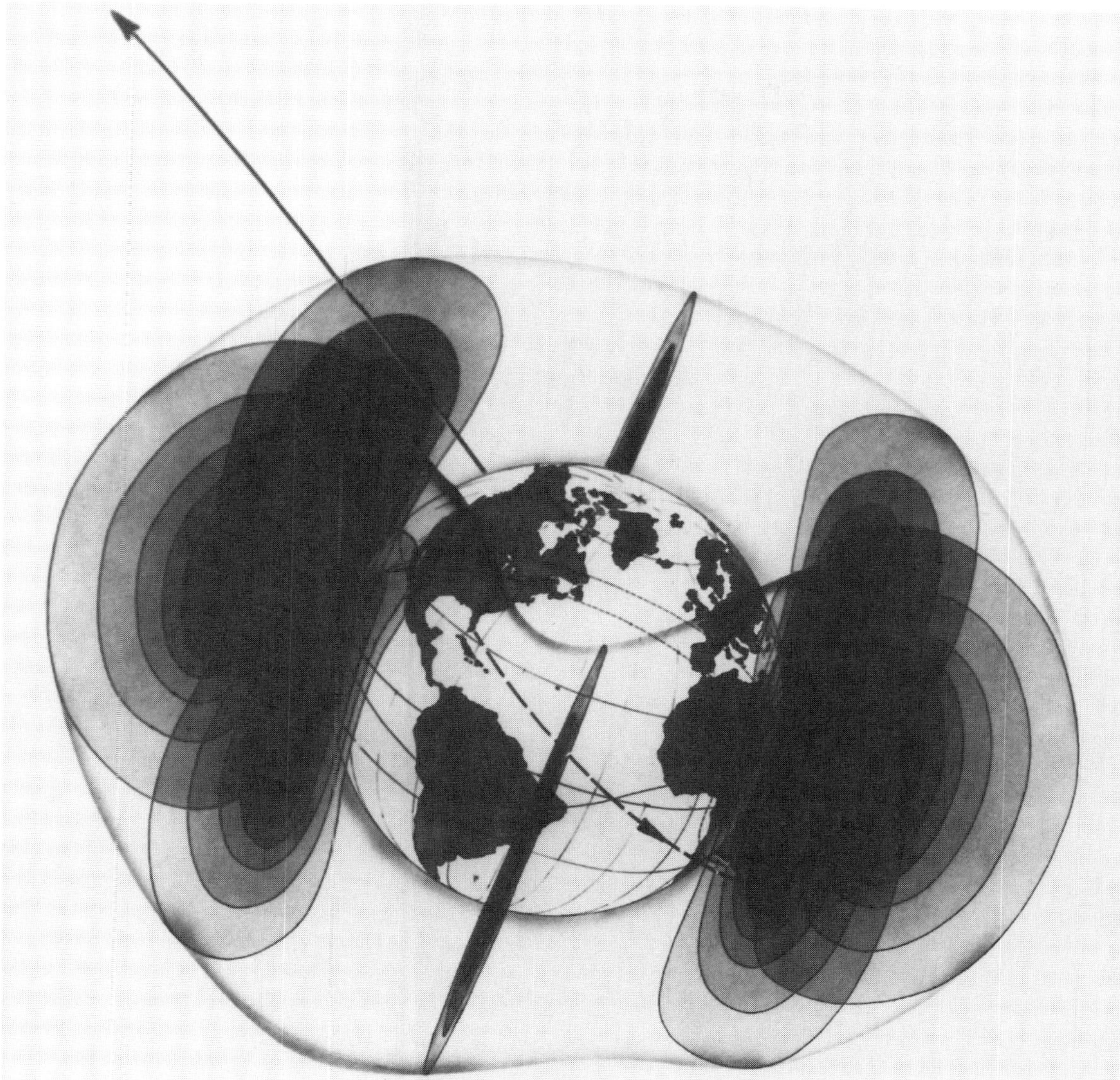

counter and causing it to register zero. Confirming his theory, the counter gave a normal reading, which showed that there were indeed cosmic rays beyond 600 miles.

This 1961 sketch of the Van Allen radiation belts shows a space vehicle trajectory plotted to avoid the greater portion of the belts by escaping along one of the relatively clear polar regions.

Discovers Van Allen belts

By May 1958 Van Allen had prepared an explanation for these results, suggesting the existence of two belts of radiation. Later research confirmed Van Allen's hypothesis and proved that the belts consist of high-velocity (high-speed) protons and electrons spiraling around Earth's magnetic lines of force. Scientists speculate that the charged particles are a result of solar flares, sudden outbursts of gases or energy from small areas on the Sun's surface. The particles are carried by solar wind (the flow of charged atomic particles that radiates from the Sun) into Earth's atmosphere, where they are trapped and cause such light spectacles as the aurora borealis at the North and South Poles. The Van Allen belts provided a new basis for investigation of the upper atmosphere.

For his discovery of the belts and other contributions to science, Van Allen has received many scientific awards and honors, including the Space Flight Award of the American Astronautical Society, the John A. Fleming Award of the American Geophysical Union, and the Elliott Cresson Medal of the Franklin Institute. Van Allen married Abigail Fithian Halsey on October 13, 1945. They have five children, Cynthia, Margot, Sarah, Thomas, and Peter.

Further Reading

Van Allen, James, ed., *Scientific Uses of Earth Satellites,* University of Michigan Press, 1956.

Van Allen, James, Richard O. Fimmel, and Eric Burgess, *Pioneer: First to Jupiter, Saturn, and Beyond,* National Aeronautics and Space Administration, 1980.

Halacy, D. S., Jr., *They Gave Their Names to Science,* Putnam, 1967.

An Wang

Born February 7, 1920
Shanghai, China
Died March 24, 1990
Boston, Massachusetts

Chinese-born American computer scientist An Wang is best known for founding Wang Laboratories, a prominent manufacturer of office word-processing and data-processing computers in the 1970s and 1980s. His early work as a computer scientist focused on the development of magnetic core memories for computers in the late 1940s and early 1950s.

Chinese-born American computer scientist An Wang was a pioneer in the manufacture of office word-processing and data-processing computers.

Goes to Harvard

An Wang was born in Shanghai, China, on February 7, 1920. His father, Yin Lu Wang, taught English in a private school. His mother's name was Zen Wan Chien. In elementary school Wang excelled in math and science, but his grades in other subjects were so poor he almost didn't graduate. In high school some of his textbooks were written in English, which would help in the future when he moved to the United States.

IMPACT

An Wang is best known for founding Wang Laboratories, a highly successful manufacturer of office word-processing and data-processing computers. During the late 1940s and early 1950s Wang conducted important work on the development of magnetic core memories for computers, ideas that were used extensively in computers until the silicon chip was invented to store memory in the late 1960s. Wang Labs became one of the giants in the computer industry in the 1970s and 1980s by selling user-friendly, inexpensive products.

After he graduated high school Wang entered the prestigious Chiao-Tung University, where he spent more time competing at table tennis than he did on his electrical engineering studies. Nonetheless, he received a bachelor of science degree in 1940. During World War II Wang assisted in the war effort by designing and building radio transmitters at the Central Radio Works. In 1945 he left China for the United States, where he enrolled at Harvard University in Cambridge, Massachusetts. He received a master of science degree in electrical engineering in 1946, then continued studying at Harvard for a Ph.D. in applied physics, which he completed in 1948.

That same year Wang met his future wife, Lorraine Chiu, who was also born in Shanghai and was a student at Wellesley College. They married in 1949 and would have three children. In April 1955 the Wangs became naturalized American citizens.

Develops magnetic core memory

Wang got his start in applied computer electronics when he went to work as a research fellow for computer pioneer Howard Aiken (see box) at the Harvard Computation Laboratory in May 1948. In 1944 Aiken had developed the Mark I, the first large American digital computer to work from a program. At Harvard Wang did his most important work in computer memories, which are essential to the development of computers as we know them today. Without memories, stored programs cannot exist, nor can programming languages or computer applications.

When Wang went to work in the Harvard Computation Lab, there were already several kinds of computer memory: magnetic drums, punched cards, vacuum tubes, electromechanical relays, mercury delay lines, and cathode-ray tubes. Each kind of memory had its disadvantages. Magnetic drums and punched cards were too slow, vacuum tubes burned out too often, and electromechanical relays were too noisy. Mercury delay lines made it difficult for users to retrieve specific bits of data in a larger data set, and cathode-ray tubes required a constant source of power or the data was lost.

Aiken wanted Wang to invent a memory that would let a computer read and record data magnetically without the mechanical movement involved in a relay or rotating drum. At first Wang was perplexed, because he found that when magnetic data was read, the process of reading the data destroyed it. But Wang soon discovered that he could use the data to rewrite the information in magnetic cores immediately after he destroyed it in the process of reading. Wang's ideas were used extensively in computers until magnetic core memories were replaced in the late 1960s by silicon computer chips. In 1955 Wang patented his ideas about reading and rewriting the information in magnetic memory cores.

Starts Wang Laboratories

Harvard preferred to sponsor basic research and decreased its work in computers when they started to develop commercially. Aware of this, in 1951 Wang founded his own company, Wang Laboratories, based in Cambridge, Massachusetts, to manufacture magnetic core memories. Since he did not have much money (only about $600), he subsisted for a while on contracts for manufacturing memories, on teaching, and on consulting. By 1953 he had a consulting contract with IBM for $1,000 a month that brought him some financial stability. In March 1956 Wang sold his patent on magnetic core memories to IBM for $500,000.

Manufactures successful products

In the early 1960s Wang Laboratories developed a popular typesetting system that would justify (align) and hyphenate text

Howard Aiken, American Computer Scientist and Inventor

A noted scientist and Harvard University professor, American computer scientist and inventor Howard Aiken (1900–1973) designed and built the Mark I calculator in the late 1930s and early 1940s. The first automatic binary computer (a calculator whose language consists of a two-digit number system representing on-off responses), the Mark I provided the groundwork for larger and more advanced computing machines. With a grant from IBM and a navy contract, Aiken began work on the Mark I with a team at IBM laboratories in Endicott, New York. An electromagnetic machine (electronically controlled, with mechanical parts), the Mark I used ordinary telephone relays that enabled electrical currents to be turned on or off. The computer consisted of thousands of relays and other components, all assembled in a 51-foot-long, 8-foot-high stainless steel and glass frame. It was completed in 1943 and installed at Harvard a year later.

The Mark I did not resemble modern computers, either in appearance or in the principles of operation. The machine had no keyboard, for instance, but was operated with approximately fourteen hundred rotary switches that had to be adjusted to set up a run. Seemingly clumsy by today's computer standards, the Mark I nevertheless was an improvement over its predecessors in terms of the speed at which it performed a host of complex mathematical calculations. Mark I had three successors, Mark II, III (Aiken's first electronic machine), and IV. Aiken never received a patent, since he disliked the idea and was known for sharing his work with others.

on a page. By 1964 the company had sales of more than $1 million for the first time. Wang then began to develop desktop calculators. One of these, the Model 300, was especially successful because it was user friendly, small, and relatively inexpensive at $1,695. By 1967 Wang's sales were up to $6.9 million per year. When Wang Labs offered to publicly sell part of its stock for sale in August of that year to raise money and eliminate some debt, the response was so enthusiastic that the value of the company soared, overnight, from $1 million to $70 million.

Produces first word processor

Wang soon realized that the future of the company was not in desktop calculators but in computers. He feared that desktop calculators would become lower-valued commodities because of increasing competition. Wang's assumption was correct; in 1971 the first pocket calculator was manufactured by Bowmar Instruments, and in the following decades the appliances became even smaller and less expensive. In response, Wang Laboratories began producing its first word processors, the Wang 1200, in 1972. Primitive by today's standards, the Wang 1200 stored data on a cassette tape and had no means of displaying text on a screen. Wang decided on some major improvements, and Wang Labs caused a sensation when it demonstrated its first CRT (cathode-ray tube) based word processor in 1976. Within two years the company had become the largest distributor of CRT systems in the world, and by 1982 sales were over $1 billion a year. That amount was tripled by 1989.

Wang died of cancer on March 24 of the following year; he was seventy years old. For his accomplishments in the computer industry he received over a dozen honorary doctorates and was named a fellow of the American Academy of Arts and Sciences.

Further Reading

Bulkeley, William M., and John R. Wilke, "Steep Slide: Filing in Chapter 11, Wang Sends Warning to High-Tech Circles," *Wall Street Journal,* August 1992, pp. A1, A6.

Cortada, James W., *Historical Dictionary of Data Processing: Biographies,* Greenwood Press, 1987.

Wang, An, with Eugene Linden, *Lessons: An Autobiography,* Addison-Wesley, 1986.

Wilke, John R., "Wang Labs Reorganization Is Cleared, Allowing Emergence from Chapter 11," *Wall Street Journal,* September 21, 1993, p. B6.

Levi Watkins Jr.

Born June 13, 1945
Parsons, Kansas

Levi Watkins Jr., has conducted research on congestive heart failure, a condition in which the heart is unable to pump out blood returned to it by the veins and thus cannot maintain an adequate blood supply to bodily tissues. He also performed the first implantation of the Automatic Implantable Defibrillator (AID), a device designed to restore normal heart rhythm during attacks of ventricular fibrillation or arrhythmia (pronounced a-RITH-mee-ah), which produce rapid, irregular heartbeats. When arrhythmia occurs, the heart is unable to pump blood and, unless corrected by devices such as the AID, the sufferer can die.

"All of a sudden I just decided teaching wasn't what I wanted to do. I wanted to be a doctor."

Decides on medical career

Watkins was born on June 13, 1945, in Parsons, Kansas, to Levi Watkins Sr., an educator, and Lillian Bernice Varnado. While he was still an infant his family moved to Montgomery, Alabama, where his father later became president of Alabama

IMPACT

Levi Watkins Jr. is best known as the first surgeon to implant the Automatic Implantable Defibrillator (AID), a device used to control irregular heartbeats in patients who suffer arrhythmias. At the time that he performed the surgery, in February 1980, coronary arrhythmias were one of the leading causes of death in the United States. Watkins has also conducted extensive research on congestive heart failure and heart disease and is a pioneer in the application of lasers to heart surgery. The first African American professor of surgery at Johns Hopkins University, Watkins actively recruits African Americans for the university medical school. In 1983 *Ebony* magazine named him one of the top fifteen African American doctors in the country.

State University. After graduating as valedictorian of his high school class in 1962, Watkins went on to study biology at Tennessee State University in Nashville. He became active in a variety of political organizations on campus and was elected president of the student government association. Although he had originally intended to follow in his father's footsteps and become a college professor, he decided during his junior year that he wanted to be a doctor.

In 1966 Watkins became the first African American student ever admitted to the Vanderbilt University School of Medicine in Nashville. He learned about his acceptance from a front-page headline in a Nashville newspaper. Watkins excelled at Vanderbilt and was later selected for membership in the Alpha Omega Alpha medical honor society. After earning a medical degree in 1970, he considered remaining at Vanderbilt for his internship. However, the chief of surgery at Vanderbilt urged him to go to Johns Hopkins University in Baltimore, Maryland, one of the foremost medical schools in the nation. He was immediately admitted for an internship in surgery.

In 1973 Watkins interrupted his surgical training to complete two years of cardiac (heart) research at Harvard Medical School, investigating the relationship between congestive heart failure and the physiology (physical and chemical reactions) of the kidney. Watkins returned to Johns Hopkins in 1975 to complete his surgical residency, and three years later became the institution's first African American chief resident in cardiac surgery. In 1978 he joined the faculty as an assistant professor of surgery and in 1991 was promoted to the rank of full professor. Again Watkins set a milestone by becoming the first African American physician to hold this position at Johns Hopkins.

Performs groundbreaking surgery

Seven months after joining the surgical team at Johns Hopkins Hospital, Watkins performed the operation that launched his career and earned him the respect and admiration of the international medical community. On February 4, 1980, he implanted the Automatic Implantable Defibrillator (AID). Watkins's patient was a fifty-four-year-old California woman who would otherwise have died of cardiac arrest.

At the time Watkins performed the operation it was estimated that 500,000 people died from arrhythmia annually, making the disorder one of the leading causes of death in the United States. Arrhythmia is an irregularity of the heartbeat caused by the presence of scar tissue in the heart or hardening of the coronary artery. The AID, invented by Michel Mirowski, the director of the coronary care unit at Sinai Hospital in Baltimore, is a small, battery-operated generator that is implanted in the patient's abdomen. One electrode leading from the AID is inserted into the right chamber of the heart; a second electrode is affixed to the tip of the heart. When the AID senses an abnormal heart rhythm, it administers mild shocks to restore the normal rhythm. The success of the device means a positive prognosis for patients who do not respond to medication for the disorder (about 25 percent). Watkins's initial AID surgical procedure was soon followed by dozens of successful implantations, and representatives of medical centers throughout the country applied to be trained for the procedure.

Pioneers laser surgery

In addition to his work with cardiac arrhythmia at Johns Hopkins, Watkins has been a pioneer in the application of lasers to heart surgery. He has also directed research on heart disease, particularly as it affects minorities, through the Minority Health Commission and Panel for Coronary Artery Bypass Surgery in Maryland. An aggressive recruiter of African American students for Johns Hopkins Medical School, he was appointed in 1979 to the university's admissions committee. In 1983 Watkins joined the national board of the Robert Wood Johnson Minority Faculty Development Program. That same year he was named one of the top fifteen African American physicians in the United States by *Ebony* magazine. In 1991 Watkins was appointed dean for postdoctoral programs and faculty development at Johns Hopkins. His other professional affiliations include the American Board of Surgery and the American Board of Thoracic Surgery.

Further Reading

Baltimore Evening Sun, April 8, 1979; August 15, 1980; May 22, 1983; February 24, 1987; April 22, 1987.

Baltimore Magazine, April 1988.

"Device Averts Heart Attacks," *Science Digest,* March 1981, p. 103.

"Finding New Breakthroughs in Heart Care," *Black Enterprise,* October 1988, p. 58.

"Young Surgeon Brings New Hope to Heart Patients," *Ebony,* January 1982, pp. 96–98, 100.

James D. Watson

Born April 6, 1928
Chicago, Illinois

James D. Watson, along with Francis Crick, determined the structure of DNA.

James D. Watson shared the 1962 Nobel Prize in physiology or medicine with **Francis Crick** (see entry) and Maurice Wilkins for determining the structure of deoxyribonucleic acid (DNA), the substance that carries genetic information from generation to generation. Watson and Crick's research ranks as one of the greatest advances in molecular biology. Indeed, more than thirty years after unraveling the structure of DNA, Watson became the director of the Human Genome Project, an attempt to map every human gene, a program made possible by his and Crick's groundbreaking work on DNA.

Considered a child prodigy

James Dewey Watson was born in Chicago, Illinois, on April 6, 1928, to James Dewey and Jean (Mitchell) Watson. Considered a child prodigy, he possessed intellectual skills that exceeded those of adults. His unusual powers of memory

IMPACT

James D. Watson is best known for unraveling the structure of deoxyribonucleic acid (DNA). He has spent most of his professional life, however, as a professor, research administrator, and public policy spokesman for research. Watson's less celebrated contribution to science is his acting as a scientific catalyst who lends support to and encourages scientists conducting various research projects at the research institute at Cold Spring Harbor, New York, of which he has been director since 1968.

led to his being one of the original Quiz Kids on the radio show of the same name. After completing high school in two years, he enrolled at the University of Chicago at the age of fifteen. Since he loved birdwatching he considered majoring in ornithology (the study of birds), but he decided to major in biology. Four years later, near the age when other young men and women were entering college, Watson graduated. While at Chicago he had developed a keen interest in genetics (the study of genes), so he went on to Indiana University at Bloomington for graduate work in that subject.

Indiana University was well known for its genetics program, which was headed by geneticist Hermann Müller, who had won a Nobel Prize, and Italian microbiologist Salvador Luria. Watson arrived a few years after a team led by Canadian-born biologist and bacteriologist Oswald Avery had uncovered evidence that deoxyribonucleic acid (DNA) played a role in passing genetic information from parent to offspring. At the time, their discovery was still not accepted by many scientists, but Watson was fascinated with the subject. Studying under Luria, he wrote his doctoral thesis on the effect of X rays on the rate of phage lysis. (A phage, or bacteriophage, is a bacterial virus. Lysis is the disintegration or bursting of cells). The biologist Max Delbrück and Luria, along with a number of others who formed "the phage group," had demonstrated that phages could exist in a number of mutant (altered) forms.

Upon earning his Ph.D. degree in 1950, shortly after his twenty-second birthday, Watson went to the University of Copenhagen in Denmark to investigate the molecular structure of proteins. The following year he met the English scientist Maurice Wilkins, who was attempting to use X rays to photograph DNA molecules. Wilkins's work sparked Watson's commitment to discovering the structure of DNA. After a year in Copenhagen, Watson received a research fellowship at the Cavendish Laboratory in Cambridge, England. At Cavendish he met Francis Crick, an English molecular biologist who was working with X-ray diffraction at the time. (X-ray diffraction is a technique for studying the crystalline structure of molecules. In this technique X rays are directed at a compound, and the subsequent scattering of the X-ray beam reflects the configuration of the molecule on a photographic plate.) Crick also became interested in DNA structure after conversations with Watson.

Intermittently over the next two years Watson and Crick theorized about DNA and worked on their model of DNA structure, eventually arriving at the correct structure by recognizing the importance of X-ray diffraction photographs produced by Wilkins and **Rosalind Franklin** (see entry) at King's College at the University of London. They were certain the answer lay in model building, and Watson in particular was impressed by the model-building method used by American chemist **Linus Pauling** (see entry) in determining the alpha-helix (coil-shaped) structure of protein. Employing the rules of chemistry, Pauling had fit together "Tinkertoy" pieces that stood for the atoms known to be present in a given molecule.

Builds DNA model with Francis Crick

Working with Tinkertoy-like models, Watson and Crick determined that DNA is shaped like a spiral staircase, or "double helix." They also used data published by Austrian-born American biochemist Erwin Chargaff on the four nucleotides (or "bases") of DNA molecules, concluding that the building blocks had to be arranged in pairs. After extensive experimentation, during which they used X-ray crystallography to take pictures of crystallized forms of DNA, they found that the double helix structure corresponded to the theoretical data produced by Wilkins, Franklin, and their colleagues. Watson and Crick were also engaged in a race with Pauling who, they feared, was building his own model of DNA. In 1953 Watson and Crick published a paper on their DNA findings in the journal *Nature* (with Watson's name

James D. Watson and Francis Crick in 1959.

appearing first, due to his victory in a coin toss). Nine years later Watson, Crick, and Wilkins shared the Nobel Prize in medicine for their work on DNA.

Becomes director of Cold Spring Harbor

After completing his research fellowship at Cavendish Laboratory, Watson spent the summer of 1953 at Cold Spring Harbor on Long Island, New York, where Delbrück had gathered a group of investigators working in the new area of molecular biology. Watson then became a research fellow in biology at the California Institute of Technology, working with Delbrück and his colleagues on problems in phage genetics. In 1955 he joined the biology department at Harvard University, where he wrote *The Molecular Biology of the Gene* (1965), the first widely used university textbook on molecular biology, which has since gone through seven editions. In 1968 Watson became director of Cold Spring Harbor, carrying out his duties there while maintaining his position at Harvard. He gave up his faculty appointment at the university in 1976, however, to assume full-time leadership of Cold Spring Harbor.

Heads Human Genome Project

In 1989 Watson was appointed director of the Human Genome Project of the National Institutes of Health (see **Walter Gilbert** entry). The ambitious task of the Human Genome Project is to map every human gene, the total of which may number up to 100,000. After less than two years, however, Watson resigned in protest over policy differences in the operation of this massive undertaking. Despite these differences, Watson continues to speak out on various issues concerning scientific research and is a strong advocate of federal support of scientific research.

In addition to sharing the Nobel Prize, Watson has received honorary degrees from numerous institutions, including the University of Chicago. He was also awarded the Presidential Medal of Freedom in 1977 by President Jimmy Carter. Watson married Elizabeth Lewis in 1968; they have two children, Rufus Robert and Duncan James. In 1993 he and Crick reunited for a fortieth anniversary celebration of their discovery of the DNA molecular structure.

Further Reading

Baldwin, Joyce, *DNA Pioneer: James Watson and the Double Helix,* Walker and Co., 1994.

Crick, Francis, *What Mad Pursuit: A Personal View of Scientific Discovery,* Basic Books, 1988.

"Happy Birthday, Double Helix," *Time,* March 15, 1993.

Sherron, Victoria, *James Watson and Francis Crick: Decoding the Secrets of DNA,* Blackbirch Press, 1995.

Watson, James D., *The Double Helix: A Personal Account of the Discovery of the Structure of DNA,* Norton, 1968.

Watson, James D., John Tooze, and David Kurtz, *The Molecular Biology of the Cell,* Benjamin, 1983.

Robert Watson-Watt

Born April 13, 1892
Brechin, Scotland

Died December 5, 1973
Inverness, Scotland

Scottish physicist Robert Watson-Watt's major contribution to science was the development of radar, the process of using radio waves to detect objects. First used successfully for military defense during World War II (1939–45), radar has been described as one of the two most significant scientific achievements—the other is nuclear weapons—resulting from research conducted during the war. Watson-Watt's discovery stemmed from his work with the British Meteorological Office on tracking thunderstorms. Following the war Watson-Watt ended his affiliation with the British government and started his own business.

Scottish physicist Robert Watson-Watt is credited with the invention of radar.

Shows early interest in radio sciences

Robert Alexander Watson-Watt was born on April 13, 1892, in Brechin, Scotland. His father, Patrick Watson Watt, a master carpenter, took his last name from his mother (Watson) and his father (Watt) and passed them both along to his son,

IMPACT

In the early days of World War II, British scientists had just completed the most advanced defense system of their day: a network of radio-signal stations that could detect the approach of German bombers, day or night, in fair or foul weather. This network was given the title "Radio Detection and Ranging," but became more commonly known as radar. Although the list of scientists who contributed to its creation is long, the name of Robert Watson-Watt usually appears at the top because his work made the development of radar possible. Since World War II nonmilitary uses of radar, such as in tracking severe weather, have become widespread.

who adopted the hyphenation of his name after he was knighted in 1942. Watson-Watt's mother was Mary Small Matthew. During his early years Watson-Watt attended the Damacre Road School and the local high school, both in Brechin. After graduation he won a scholarship to University College at the University of Dundee, where he pursued a degree in electrical engineering and was introduced to the field of wireless telegraphy (radio). His fascination with the subject would eventually guide and motivate nearly all of the research he conducted throughout his life.

In 1912 Watson-Watt earned a bachelor of science degree in electrical engineering. He was then appointed assistant professor of physics at University College, a post he held only until the outbreak of World War I in 1914. Although his application for a job with the British War Office was denied, he was hired by the government Meteorological Office. While working for this weather agency Watson-Watt proposed a method for tracking severe weather patterns by triangulation with radio waves. Triangulation involves determining the network of triangles into which Earth's atmosphere is divided. Radio waves are electromagnetic waves with various radio frequencies, or electronic charges per second. The development of this technique was slow, however, because the technology required to perform related experiments had not yet been perfected.

Directs research on thunderstorms

At the conclusion of the war Watson-Watt enrolled at the University of London, where he received a bachelor of science degree in physics in 1919. Still employed by the Meteorological Office, he transferred to a field observing station at Ditton Park in Slough, England. In 1927 that station was combined with a nearby facility of the National Physical Laboratory to form a single unit called the Radio Research Station, under the authority of the Department of Scientific and Industrial Research. Watson-Watt was named head of the new station, and over the next decade he directed research on the radio location of thunderstorms, the detection of radio signals, and studies of the atmosphere. In connection with this work Watson-Watt also proposed the name "ionosphere" for the layer of the atmosphere discovered in 1924 by English physicist Edward Appleton (see box). The ionosphere is composed of gases that have been converted to ions (charged subatomic particles) and aids the reflection of radio waves.

Invents radar

Watson-Watt's most notable accomplishment came in 1935. An official with the Air Ministry had asked if it would be possible to concentrate radio waves in such a way that they could be used to destroy enemy aircraft. Although he and his assistant, A. F. Wilkins, pointed out the impossibility of creating such a device, Watson-Watt drafted a memo to the Air Ministry in which he explained how radio waves could at least be used for detecting aircraft. A beam of radio waves can be sent in the direction of a target object, he said. Then the distance to the object can be calculated by determining the length of time required for the beam to reflect off the object and return to its source. Watson-Watt's memo was dated February 12, 1935, a date he later identified in his autobiography, *Three Steps to Victory: A Personal Account by Radar's Greatest Pioneer,* as being "the birth of radar and as being in fact the invention of radar." The term radar, coined in the United States, is an acronym for "Radio Detection and Ranging."

Draws upon earlier discoveries

While Watson-Watt was the first to suggest how to use radio waves as an object-tracking technique, several important

Edward Appleton, English Physicist

Robert Watson-Watt proposed the name "ionosphere" for the layer of the atmosphere discovered in 1924 by English physicist Edward Appleton (1892–1965). Appleton's atmospheric research was inspired by the first long-distance radio messages transmitted across the Atlantic Ocean by **Guglielmo Marconi** (see entry) in 1901. After Marconi's successful transmission two physicists, Oliver Heaviside and Arthur E. Kennelly, postulated the existence of an atmospheric band of gases that have been converted to ions (charged subatomic particles). These bands aid the reflection of radio waves (a form of electromagnetic radiation). More than twenty years later Appleton, then a professor of physics at King's College at the University of London, verified that hypothesis. In 1924, assisted by graduate student Miles Barnett, Appleton designed and carried out a series of experiments to determine how radio waves are transmitted in the atmosphere.

Appleton and Barnett were able to convince the British Broadcasting Corporation (BBC) to allow them to use the network's radio signals outside of scheduled broadcasting hours. On a regular basis they varied the BBC signal frequency. They were looking for points at which interference occurred between direct waves from the atmosphere. Their results indicated the existence of a reflecting layer in the atmosphere approximately 60 miles (100 kilometers) above Earth's surface. Known as the E layer, it occurs during daylight hours and is capable of reflecting shortwave frequencies. (A shortwave is a radio wave having a wavelength between 10 and 100 meters.)

Appleton continued to conduct research on the atmosphere throughout his life. In 1926 he discovered a second reflecting layer above the E layer. Called the F layer (or Appleton layer), it is the most highly ionized layer of the ionosphere. The F layer occurs at night within the F region, the highest region in the ionosphere, from 80 miles to more than 300 miles (130 to 500 kilometers) above Earth's surface. Appleton's research also led him to discover a number of important characteristics of the reflecting layers. In 1927, for example, his observation of a solar eclipse (the complete or partial obstruction of the Sun by the Moon) convinced him that the existence of ionized layers is the consequence of solar radiation bombarding Earth's atmosphere. For his work on atmospheric structure Appleton was awarded the 1947 Nobel Prize in physics.

discoveries were instrumental in the invention of the radar system. For example, in 1922 the cathode-ray tube, or CRT (a vacuum tube in which a beam of electrons is projected on a fluorescent screen), which would be used for visualizing the returning radar signal, became available. In 1936 pulsed radar (the release of radio waves in regular, brief spurts) replaced continuous-wave emitters. Whereas the old system could detect only the presence of an object, the pulsed signal could pinpoint its location. Finally, in 1939 the practical microwave transmitter (a device for emitting short electromagnetic waves, between about one millimeter and one meter in wavelength) was constructed. This system allowed the radar locator to operate through cloud and fog.

Britain builds radar defense system

The British government immediately initiated a secret program for the implementation of Watson-Watt's idea, constructing a radar defense shield. Throughout Britain a network of radio-signal stations were built for detection of German bombers, day or night, in fair or foul weather. Most military authorities agree that the radar system was the deciding factor in repulsing Germany's air invasions during the Battle of Britain in 1940. Ironically, German scientists had reportedly been working on a radar location system of their own, but German leaders, including Adolf Hitler and air force minister Hermann Göring, saw it as being useful only in defensive situations. Since Hitler and Göring assumed that Germany would never have to defend itself against invaders, the development of radar was considered a low priority. Watson-Watt was invited to the United States in 1941 to help Americans develop their own radar system.

Starts his own company

Throughout his life Watson-Watt had been interested in important social issues outside the field of science. From 1929 to 1936, for example, he was an active member and twice an officer of the Institute of Professional Civil Servants. He was also a fellow (member) and treasurer of the Institute of Physics, a fellow of the Royal Society, and the recipient of numerous honorary degrees. In 1946 he formed a private consulting company, Sir Robert Watson-Watt and Partners, and consulted with a number of agencies within the British government. (Sir is the title Watson-Watt received after being knighted in 1942.)

Watson-Watt was married three times between 1916 and 1966. His last wife, Dame Katherine Trefusis-Forbes, formerly head of the Women's Royal Air Force, died in 1971. Details regarding the last two decades of Watson-Watt's life are not certain. According to an account in *Biographical Memoirs of the Fellows of the Royal Society,* he lived mainly in Canada and the United States, working from time to time as a freelance scientific adviser. Watson-Watt died in Inverness, Scotland, on December 5, 1973.

Further Reading

Biographical Memoirs of Fellows of the Royal Society, Volume 21, Royal Society (London), 1975, pp. 549–68.

Dictionary of Scientific Biography, Scribner, 1982, pp. 977–78.

Watson-Watt, Robert, *Man's Means to His End,* Heinemann, 1962.

Watson-Watt, Robert, *Three Steps to Victory: A Personal Account by Radar's Greatest Pioneer* (autobiography), Odhams, 1957, abridged version published in the United States as *The Pulse of Radar,* Dial, 1959.

James Watt

Born January 19, 1736
Greenock, Scotland

Died August 19, 1819
Birmingham, England

Scottish engineer and inventor James Watt developed the steam engine, a machine that was instrumental to the rise of the Industrial Revolution of the late eighteenth and early nineteenth centuries in Europe. His improvements to existing steam engine technology produced a faster, more efficient machine that could more easily power heavier machinery. Throughout his life Watt experimented and improved on his engine, refining systems that provide the basis for many industrial engines used today.

Scottish engineer and inventor James Watt invented the steam engine, a machine that revolutionized industry.

Early life

James Watt was born on January 19, 1736, in the small town of Greenock, Scotland. His father made a living however he could, working variously as a merchant, a carpenter, and a government administrator. Young Watt frequently assisted his father as a carpenter and from an early age enjoyed carpenter's tools more than books. He was a physically weak child who

IMPACT

James Watt's many refinements to the steam engine fueled the Industrial Revolution in Europe and revolutionized methods of manufacturing worldwide. After creating a more energy-efficient steam engine in 1775, he turned it upside down, making a "topsy turvy" model that could operate a pump directly without the need for a large and cumbersome rocker arm. He found a way to turn the engine on its side so that it could be used in other factory activities. He also created the first steam valve that would allow steam to apply force alternately on either side of the piston. Watt later invented a series of gears called the "sun and planet motion" that replaced the push-pull motion of the engine with the circular motion needed to power mills. For the first time mills and factories were not limited to locations near streams or windy plains.

experienced long periods of illness, so he did not begin his education at the usual age. Even when he did enroll, illness frequently kept him from class. Watt demonstrated a keen ability to learn on his own, however, and as he played with carpenter's tools he learned and applied mathematical concepts.

In his teens Watt was sent to a commercial school, where he studied Latin, Greek, and mathematics. At home he experimented with chemicals and created new tools and by the age of fifteen had found a way to make a simple electric motor. Watt spent a year at the University of Glasgow in Scotland, where he studied to become a mathematical instrument maker. He then moved to London and worked for a man named John Morgan, making accurate quadrants (tools for measuring the altitude of stars) and parallel rulers.

"Inventor" of the steam engine

Watt returned to Scotland when he was twenty. He took a job at the University of Glasgow, cleaning and repairing scientific instruments. It was at Glasgow that he first became interested in steam power. Though Watt is often credited with the "invention" of the steam engine—and he certainly deserves credit for making steam power popular and for adapting it to many uses—earlier inventors had experimented with steam power. In fact, steam engines were already popular in one of England's largest industries, coal mining. They were used to power pumps to keep the mines from flooding, and they were economical because low-grade coal, which they used for fuel, was readily available.

The use of steam to power machines initially depended on the ability to create a vacuum. For instance, the engines used to pump water from the mines were based on expansion and suction effects achieved by generating and then condensing steam to create a vacuum. The process was begun when water was heated in the boiler chamber until its steam filled the chamber. This created an expansion that forced out any water or air inside the chamber. A valve was then closed, after which cold water was sprayed over the chamber; this chilled and condensed the steam inside to form the vacuum. When the valves were reopened, water was forced up from the mine, and the process could then be repeated.

British engineer Thomas Newcomen introduced a steam pump in 1712 that operated on the same principle, but he increased efficiency by setting a moving piston inside a cylinder, a technique still used today. A cylinder, a long, thin, closed chamber separate from the boiler, replaced the large, open boiler chamber; and a piston, a sliding piece that fits in the cylinder, was used to create motion instead of a vacuum. Newcomen's design was significant in its capacity to produce motion to power a machine. The motion created allowed the engine to sustain its own movement; earlier steam-powered machines required that their water and steam valves be constantly monitored and manipulated.

But Newcomen's engines had a number of weaknesses, which Watt soon recognized and set about to improve. Newcomen's engine is called an atmospheric engine because its motion relies on atmospheric pressure rather than on counterweights or the force of steam itself. Steam was admitted to the underside of a piston, then, at a determined point in the cycle, a jet of water was injected to the same area to condense the steam and create a vacuum. The atmospheric pressure in the open end of the cylinder then forced the piston back down, producing a power stroke. Newcomen's machine was capable of twelve strokes per minute.

Watt determined that Newcomen's engine wasted time by cooling the piston chamber during every cycle. Newcomen's engine—as did other early steam pumps—injected steam on only one side of the cylinder and employed atmospheric pressure and vacuums to reset the piston in position. These earlier engines suffered in efficiency because heat was lost during the

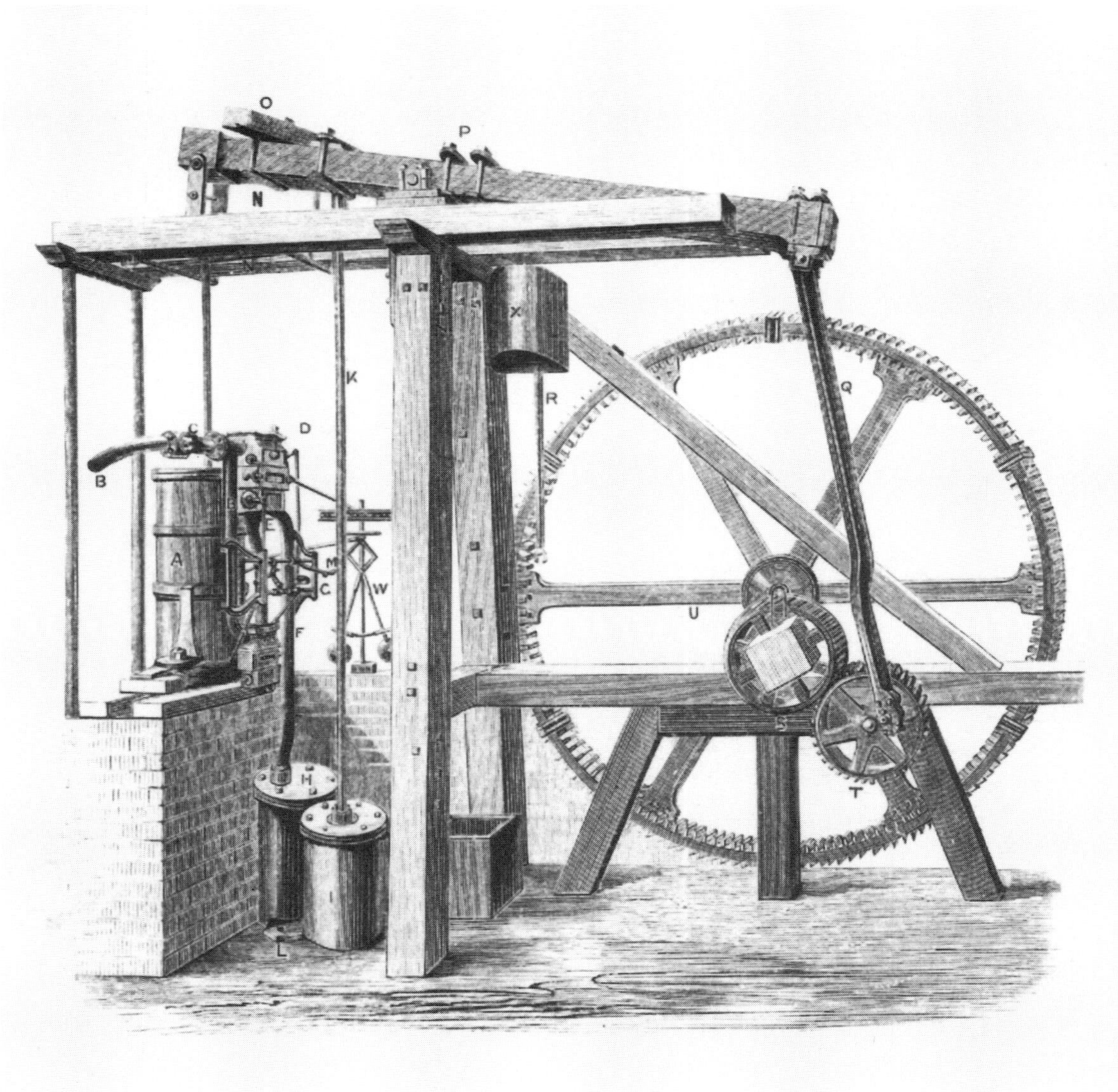

Schematic drawing of James Watt's first rotary steam engine.

procedure, which made reheating necessary at the beginning of each cycle. In 1769 Watt obtained a patent on an engine design that included sealing the engine's cylinder and installing steam valves on both ends, innovations that would leave the steam chamber always hot and always ready for a new batch of steam, thereby allowing a large pump to be driven back and forth faster than Newcomen's twelve strokes a minute.

But Watt found it difficult to seal the cylinders because those available were still bored (drilled) as they had been in Newcomen's time, sixty years earlier, and would not seal tightly enough for Watt's purpose. To address this problem, Watt entered into a partnership in 1773 with businessman Matthew Boulton (see box), who financed and organized a search for techniques to manufacture a well-sealed cylinder. Two years later English industrialist and inventor John Wilkinson perfected a boring machine that could drill cylinders with unprecedented uniformity.

Watt's refined steam engine design used one-third less fuel than a comparable Newcomen engine. Watt and Boulton began to market these engines, and in 1775 several of them were sold to operate water pumps and, during the next year, two more were installed in coal mines.

Initiates the Industrial Revolution

Over the next fifteen years Watt continued to experiment with and improve upon steam engine technology. He introduced the centrifugal governor, which became known as the Watt governor, a device that could control and adjust steam output and engine speed. He devised a new type of condenser that used a system of tubes instead of one large chamber, and an air pump that maintained a vacuum in the condenser. But his most impressive innovation was attaching a flywheel to the engine. Flywheels, heavy wheels that control speed in machinery with which they revolve, accomplished two tasks: they allowed the engine to run more smoothly by creating a more constant load, and they converted the conventional back-and-forth power stroke into a circular (rotary) motion that could be adapted more readily to power machinery. It was this step perhaps more than any other that led the steam engine to the forefront of Europe's Industrial Revolution. It was used to pump bellows for blast furnaces, to power huge hammers for shaping and strengthening forged metals, and to turn machinery at textile mills. For the first time mills and factories were not limited to locations near streams or windy plains.

Matthew Boulton, English Manufacturer and Engineer

In 1775 English manufacturer and engineer Matthew Boulton (1728–1809) entered into partnership with James Watt to build an improved steam engine. During the next decade their engines were to become prime forces behind the Industrial Revolution in Europe, and Boulton's fame grew as he became the leading English manufacturer, the man who brought Watt's genius to fruition by means of his manufacturing expertise. "I sell here what all the world desires—power," Boulton proclaimed, as quoted by the Scottish writer James Boswell. In his later years Boulton patented a steam-driven coin press that restored public confidence in the English monetary system and forced counterfeiters out of business. A founding member of the prestigious Lunar Society of Birmingham, at the time of his death Boulton ranked as one of the most prominent scientific figures of the era.

By 1790 Watt's engines had nearly replaced Newcomen's engines, which were, for the most part, decades old. Ten years later, at age sixty-four, Watt sold his interests in his company to his son, James. Despite his success, Watt is said to have been melancholy and withdrawn by nature. He had warm friendships but hated to meet face-to-face with anyone in business negotiations. In his old age, however, he demonstrated a desire to share his knowledge and reminisce about his past successes. The watt, a metric unit of power, is named for him.

Watt died on August 19, 1819, near Birmingham, England. He is buried beside his former business partner, Matthew Boulton.

Further Reading

Crane, William D, *The Man Who Transformed the World: James Watt,* Messner, 1963.

Holt, L. T. C., *Thomas Newcomen: The Prehistory of the Steam Engine,* Dawlish MacDonald, 1963.

Smiles, S., *Lives of Boulton and Watt,* Murray, 1865.

Storer, J. D., *A Simple History of the Steam Engine,* John Baker, 1969.

Alfred Wegener

Born November 1, 1880
Berlin, Germany
Died November 1, 1930
Greenland

German meteorologist and geophysicist Alfred Wegener gained renown for proposing the idea of continental drift, even though his theory that the earth's continents were all once connected was not accepted during his lifetime. In the 1960s, decades after his death, however, oceanic data convinced scientists that continents do indeed move. As a result, the theory of continental drift that Wegener had proposed in 1912 became the well-established foundation for the theory of plate tectonics, which states that sections of the earth's crust do indeed move. Wegener died during one of his scientific trips to Greenland.

German meteorologist and geophysicist Alfred Wegener originated the theory of continental drift.

Goes to Greenland

Alfred Wegener was born in Berlin, Germany, on November 1, 1880, to Richard and Anna Wegener. He studied at the universities in Heidelberg, Innsbruck, and Berlin, receiving a doctorate in astronomy from the University of

Berlin in 1905. Wegener's thesis involved conversion of a thirteenth-century set of astronomical tables into decimal notation. Thereafter he abandoned astronomy in favor of meteorology, the study of the earth's atmosphere and its phenomenon, especially weather and weather forecasting. Fascinated with this new science, Wegener carried out experiments with kites and balloons. In 1906 he and his brother Kurt set a world record in an international balloon contest by flying for fifty-two hours straight. That same year Wegener fulfilled his lifelong dream of going to Greenland. He was chosen as the official meteorologist for a two-year Danish expedition to the large North American island, most of which lies in the Arctic Circle. It was the first of four trips he would take to Greenland. In 1912, when he returned with an expedition to study glaciology (the study of glaciers) and climatology (the study of the earth's climate), he accomplished the longest crossing of the ice cap ever made on foot.

Begins atmospheric studies

In 1908 Wegener accepted a job teaching meteorology at the University of Marburg in Germany. Three years later he published a textbook on the thermodynamics (the relationships between heat and other energy forms) of the atmosphere, which initiated the modern theory on the origins of precipitation. The following year Wegener married Else Köppen, the daughter of Wladimir Köppen, the "Grand Old Man of Meteorology" in Germany. During World War I (1914–18) Wegener served as a junior military officer and was wounded twice. After the war he succeeded his father-in-law as director of the meteorological research department of the Marine Observatory near Hamburg. There he conducted experiments to reproduce lunar craters by hurling projectiles (objects) at various ground substances, demonstrating that the craters were probably caused by impact rather than volcanic movement. He also continued to analyze the data from Greenland, observe meteorological phenomena, and develop his earlier ideas on the origin of the continents and the oceans.

Presents idea of continental drift

Wegener had first thought of the idea of continental drift in late 1910 while looking at a world map in an atlas. He noticed that the east coast of South America matched like a puzzle piece with the west coast of Africa, but dismissed the idea of drifting continents as improbable. The next year, however, he came across a list of sources arguing that a land bridge must have connected the two continents at one time, since similar fossils from the same time period appeared in both Africa and Brazil. Wegener immediately began to search out fossil evidence to support the idea of drifting continents. Within a few months he presented his hypothesis in two public forums.

IMPACT

Alfred Wegener devised the first coherent and logical argument for continental drift. During his lifetime, however, he could not find an adequate mechanism to explain that the continents had once been connected but later drifted apart. Finally, in the 1960s, decades after his death, scientists discovered the theory of seafloor spreading. Here was the mechanism by which Wegener's continents could drift: the ocean floor was constantly regenerating itself. Eventually, Wegener's hypothesis revolutionized geology, becoming a cornerstone for modern views of the earth's history.

Wegener then published an extended account of his idea in 1915. Titled *Die Entstehung der Kontinente und Ozeane* ("The Origin of Continents and Oceans"), by 1924 the third edition of this work had been translated into five languages and was widely read for the first time. The first English translation correctly referred to the idea of "continental displacement," as Wegener had termed it. The name "continental drift" was coined later.

Wegener was not the first to suggest the idea of continental drift. In 1620, for instance, English philosopher and author Francis Bacon noted the physical similarities between the American and African coasts. However, Wegener presented the first coherent and logical argument for continental drift that was also supported by concrete evidence. He proposed that a huge supercontinent had once existed, which he named Pangaea, meaning "all land." He suggested that Pangaea was surrounded by a supersea, Panthalassa, and that two hundred million years ago, in the Mesozoic period, Pangaea began to separate into continents that moved away from each other. The Americas drifted westward from Europe and Africa, forming

Harry Hammond Hess, American Geologist

In the 1960s American geologist Harry Hammond Hess (1906–1969) confirmed Alfred Wegener's theory of continental drift with his hypothesis of seafloor spreading. A Princeton University professor who was an officer in the navy reserve, during World War II Hess was called on to take command of an attack transport ship. Taking full advantage of the ship's equipment for sounding the ocean floor, Hess mapped and sampled a large part of the Pacific Ocean, discovering in the process the underwater flat-topped seamounts (active or extinct volcanoes) that he named guyots, in honor of A. H. Guyot, the first professor of geology at Princeton. The origin of guyots was puzzling, for they were flat on top as if they had been eroded at the ocean surface, yet they were two kilometers below sea level. It was later determined that guyots are extinct volcanos that were once above sea level and later sunk below the surface.

After the war ended Hess continued to study guyots as well as mid-oceanic ridges, breaks in the earth's crust that run down the centers of the Atlantic and Pacific Oceans like an underwater backbone. Hess learned that at the mid-oceanic ridges, or the "seams" between the earth's plates, molten rock (magma) constantly rises toward the surface to form new oceanic crust and pushes the existing seafloor out of its way, a process Hess labeled seafloor spreading. Meanwhile, making room for new crust, old crust constantly descends into deep depressions, or ocean trenches, where it is destroyed. With the concept of seafloor spreading Hess suggested a mechanism by which continents could move away from each other without tearing up a rigid seafloor. His hypothesis gave geologists their first clue that drifting continents are carried passively on the spreading seafloor.

the Atlantic Ocean. India moved east from Africa, and Australia split away from Antarctica and moved toward the equator.

Contradicts current views

Wegener's theory contradicted the accepted view of the earth in his day. Other geologists believed that the earth was still cooling and contracting from a mass of molten rock. They said that lighter rocks such as granite (termed "sial") moved toward the surface. Underlying these rocks were denser rocks such as basalt ("sima"). Mountain ranges, they believed, were produced when the rock cooled and contracted, like wrinkles appearing on drying fruit. To these scientists, the continents and the ocean basins were permanent, so it seemed impossible for continents to move through the ocean rocks.

Wegener instead proposed that the lighter sial that made up continents could move horizontally through the oceanic sima. If the continents can rise up vertically, he argued, they must be able to move horizontally as well, as long as sufficient force is provided. Thus the Rocky Mountains and the Andes, on the western edges of the Americas, were formed by the resistance of the sima layer to the continents plowing through them. Island arcs like Japan and the West Indies were fragments left behind in the wake of these giant drifting continents.

Shows fossil evidence

Wegener's strongest evidence was the similarities of rocks, animals, and plants on both sides of the Atlantic. He pointed to the fossils of several reptiles and plant life that were known only in Africa and South America, and to the fact that the distribution of some living animals was hard to explain unless the continents had once been connected. Scientists had previously explained these in terms of a land bridge that had once connected the continents and then sunk into the ocean. Wegener argued that this was impossible because if the bridge were made of sial, it could not simply sink and disappear.

Although Wegener did not believe in a land bridge, he could not find an adequate explanation for continental drift. He suggested two theories, which were both later disproved. One was *Pohlflucht,* or "flight from the poles," to explain why continents seemed to drift towards the equator. *Pohlflucht,* also known as the Eötvös force, came from the fact that the earth is an oblate spheroid (a sphere that is slightly flattened at the poles and bulging at the equator). Next, to explain the westward movement of the Americas, Wegener suggested that some kind of tidal force must be doing the work.

Scorned by other scientists

Wegener's hypothesis was received with ridicule. For decades other geologists scoffed at the idea of drifting continents. Some scientists did support him, but there was not enough geological evidence to prove beyond a doubt that he was essentially right. His first critic was his father-in-law, Köppen, who apparently wanted Wegener to stay in meteorology and not wander into an unknown area like geophysics, the study of the physics of the earth and its environment. In 1912, at Wegener's first lecture in Frankfurt, some geologists were apparently indignant at the very notion of continental drift. In 1922, when *The Origin of Continents and Oceans* first appeared in English, it was blasted in a critical review and at a scientific meeting. Consequently, continental drift provoked an international debate, with scientists positioning themselves on both sides.

Wegener's critics had plenty of ammunition. It was soon shown that *Pohlflucht* and tidal forces were about one-millionth less powerful than they needed to be to move continents. As a result, Wegener's evidence was thought to be inconclusive. In 1928, at a meeting of fourteen eminent geologists, seven opposed it, five supported it without reservation, and two supported it with reservations. From then until after World War II (1939–45) the subject was put on the back burner of scientific debate. The only major advocate of the theory was South African geologist Alex du Toit, a vigorous defender of continental drift. In 1937 he proposed that instead of Pangaea there were two supercontinents, Laurasia in the northern hemisphere and Gondwanaland in the south.

Wegener had finally been given a professorship in meteorology and geophysics at the University of Graz in 1926. Four years later he sailed from Copenhagen to Greenland as leader of a major expedition. On November 1, 1930, he and the party celebrated his fiftieth birthday at a camp in the center of the Greenland ice cap. Wegener headed for the west coast that day and apparently died of heart failure. His body was later found about halfway between the two camps.

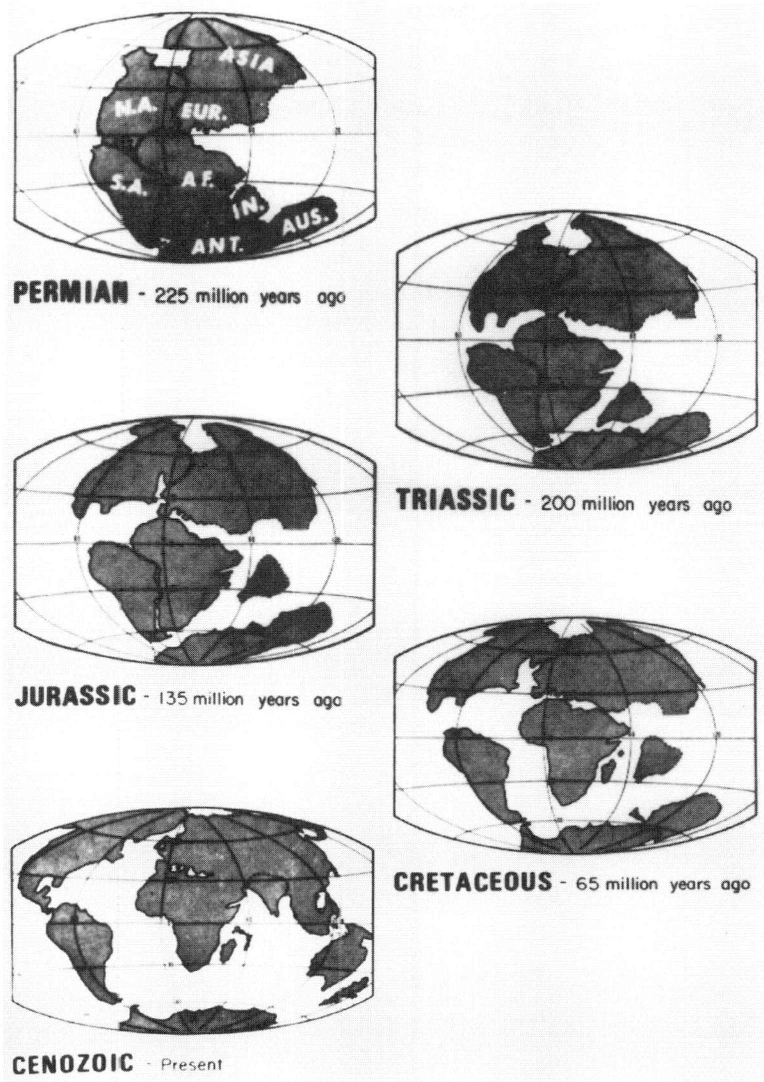

Map showing the sequence of continental drift by geological ages.

Theory leads to plate tectonics

After World War II—following Wegener's death—other geologists began to uncover clues that eventually led to a new era in geology sparked by the theory of plate tectonics. The development of paleomagnetism, the science that deals with the magnetism of rocks, in the early 1950s demonstrated that rocks in different continents appear to have different directions of magnetization, as if continents had drifted apart from each other. In addition, oceanographers began to map the ocean floor to determine its origin; they learned it was not a fixed glob of sima at all.

In 1960 American geologist Harry Hammond Hess (see box) proposed the theory of seafloor spreading: that the ocean floor is constantly being created at underwater ridges in the middle of the oceans, spreading outward, and being consumed in trenches underneath the continents. By the mid-1960s new data on magnetic anomalies (irregularities) in the Pacific Ocean revealed that seafloor spreading did indeed occur. Here was the means by which Wegener's continents could drift: the ocean floor was constantly regenerating itself. By the end of the 1960s continental drift had begun to be accepted by the entire earth science community. Although it had taken half a century, Wegener's hypothesis became the foundation for a revolution among geologists and a cornerstone for modern views of the earth's history.

Further Reading

Hallam, Anthony, *Great Geological Controversies,* Oxford University Press, 1983.

Hallam, Anthony, *A Revolution in the Earth Sciences: From Continental Drift to Plate Tectonics,* Clarendon Press, 1973.

LeGrand, H. E., *Drifting Continents and Shifting Theories,* Cambridge University Press, 1988.

Marvin, Ursula B., *Continental Drift: The Evolution of a Concept,* Smithsonian Institution Press, 1973.

Norbert Wiener

Born November 26, 1894
Columbia, Missouri
Died March 18, 1964
Stockholm, Sweden

One of the most original thinkers of his time, American mathematician Norbert Wiener was the creator of cybernetics. Cybernetics is the study of "intelligent systems" that interact with their environment, learn from their actions, and make decisions that help them adapt to the environment or achieve some specified goal. These systems may be human beings, animals, or machines. Wiener wrote for a variety of popular magazines as well as technical journals, and he was not reluctant to express unpopular political views. His intellectual originality led him down paths that subsequent generations have followed.

American mathematician Norbert Wiener was the originator of cybernetics.

Graduates from college at age sixteen

Norbert Wiener was born in Columbia, Missouri, on November 26, 1894, to Leo Wiener and Bertha Kahn Wiener. He had an unusual childhood. His father, a professor at Harvard University in Cambridge, Massachusetts, was a strict

IMPACT

American mathematician Norbert Wiener is considered the creator of cybernetics. Derived from the Greek word *kyberetes,* meaning "steersman" or "helmsman," cybernetics is the study of communication and feedback control in machines and humans. Cybernetics analyzes the ability of humans, animals, and some machines to respond or make adjustments based upon input from the environment. It is this process of response or adjustment that is called "feedback" or "automatic control." Cybernetics is most commonly used today in robotics and artificial intelligence, the science that attempts to imitate human intelligence with computers.

taskmaster who pushed his son toward success. Consequently Wiener was able to read and write by the age of three, and he graduated from high school when he was eleven. Immediately after graduation he entered Tufts College in Medford, Massachusetts, but he had a difficult time choosing a major field of study. He transferred to Harvard to study zoology, but that subject did not suit him, so he tried majoring in philosophy at Cornell University in Ithaca, New York. Wiener soon lost interest in philosophy and, at age eighteen, ended up receiving a Ph.D. in mathematics from Harvard.

Studies with great mathematicians

After receiving his degree Wiener earned a traveling fellowship that enabled him to study in Europe. Nevertheless his father, still supervising his career, wrote to the prominent British mathematical theorist **Bertrand Russell** (see entry) on Wiener's behalf. Beginning in 1913 Wiener studied for a year in England with Russell, at the same time taking courses from the British analyst G. H. Hardy, who would later influence his ideas. The following year Wiener went to the University of Göttingen in Germany, where he studied with other important mathematicians. When he returned to the United States in 1915, he was unsure about his future plans. He wrote articles for the *Encyclopedia Americana* and took a variety of teaching jobs until the United States entered World War I (1914–18).

An enthusiastic patriot, Wiener tried to enlist in the military, but he was rejected because of his poor eyesight. He then joined a group of scientists and engineers at the Aberdeen Proving Ground in Maryland, where he met Oswald Veblen, one of the leading mathematicians in the United States.

Although Wiener did not pursue Veblen's field of research, Veblen's work in producing successful results for the military impressed him more than mere academic accomplishment.

In 1919 Wiener was appointed to the faculty of the Massachusetts Institute of Technology (MIT), where he would remain for forty-one years. At that time mathematics was not a particularly strong area at MIT, but his position assured him of continued contact with engineers and physicists. As a result, Wiener pursued an ongoing interest in the relationship between mathematics and the physical world. By 1932 he had advanced through the ranks of MIT, achieving the status of full professor.

Creates cybernetics

During World War II (1939–45) Wiener developed radar and missile guidance systems, utilizing his knowledge of automatic controls and information feedback. (Feedback is the ability of machines, as well as humans and animals, to respond to or make adjustments based upon input from the environment.) After the war he began a deeper study into these subjects, particularly the similarities between the animal nervous system and the feedback mechanisms employed in early computer systems. Wiener's study comparing the animal nervous system and feedback mechanisms led to what is now called cybernetics. Cybernetics is the study of "intelligent systems" —whether these systems are humans, animals, or some machines—that interact with their environment, learn from their actions, and make decisions that help them adapt to the environment or to achieve some specified goal. According to Wiener, the central idea of cybernetics is control through feedback. One simple example of this is a thermostat that measures the temperature of its environment (feedback) and accordingly adjusts its action by turning the furnace on or off (control).

In 1948 Wiener summarized his findings in a book titled *Cybernetics,* the title taken from the Greek term *kyberetes,* meaning "steersman." Popular both within and outside the scientific community, the book served to create a new realm of

John von Neumann, Hungarian-born American Mathematician

Hungarian-born American mathematician John von Neumann (1903–1957) was notable for his work in the early development of high-speed digital computers and stored-information programs that are now used in nearly all computer applications. Along with Norbert Wiener, von Neumann is also considered a pioneer in the theory of cybernetics, although his role was not recognized until after his death. Von Neumann and economist Oskar Morgenstern collaborated on the 1944 book *The Theory of Games and Economic Behavior,* which brought about a revolution in the social sciences. In *The Theory of Games* von Neumann and Morgenstern argued that theories of mathematics developed for the physical sciences were not adequate for economics, which, after all, involves human action based on choice and chance. Proposing a different mathematical approach for the social sciences, von Neumann developed "game theory."

Game theory is a branch of mathematics concerned with the analysis of conflict situations. It involves determining a strategy for a given situation and the costs and benefits of using that chosen strategy. According to von Neumann, such "games" can be found in economics, military science, politics, or any other social process. Game theory assumes, however, that all participants act rationally at all times to maximize the outcome of the "game" for themselves. It also assumes that all participants are able to rank-order all possible outcomes without making mistakes. Although von Neumann's theory has been criticized because it does not allow for human failings, it nevertheless has been widely used in complex decision-making processes.

automatically controlled machines. Although cybernetics is most commonly implemented in robotics and artificial intelligence (the science that attempts to imitate human intelligence with computers), it has also been used in several other areas. For example, philosophers, mathematicians, and social scientists have used cybernetics to ask and answer basic questions about perception, knowledge, thinking, language, and organization.

Popularizes theory of cybernetics

Wiener treated cybernetics as a branch of mathematics with its own terms, like "signal," "noise," and "information." Another prominent figure in this area was Hungarian-American mathematician John von Neumann (see box), who did extensive work in the computer field and game theory and whose contribution to cybernetics was realized only after his death. Following the publication of *Cybernetics,* Wiener continued to lecture on the advantages of cybernetics in the workplace as well as warn of its dangers. In 1950 he published *Human Use of Human Beings,* cautioning against the dangers of exploiting the computer's potential. During his later years he also wrote for a wide variety of magazines, including the *Atlantic, Nation,* the *New Republic,* and *Colliers.* His two-volume autobiography, *Ex-Prodigy* and *I Am a Mathematician,* were published in 1953 and 1956, respectively.

Wiener married Margaret Engemann in 1926, and they had two daughters, Barbara and Peggy. In 1960 Wiener retired from MIT, and in 1964 he was awarded the National Medal of Science. Two months later Wiener died while traveling through Stockholm, Sweden.

Further Reading

Dictionary of Scientific Biography, Scribner, Volume 14, 1970–78, pp. 344–47.

Heims, Steve J., *John von Neumann and Norbert Wiener,* MIT Press, 1980.

Masani, P. R., *Norbert Wiener,* Birkhäuser, 1990.

James S. Williamson

Born November 30, 1949
Williston, North Dakota

"Solar energy was used in space long before it was on the ground."

Solar power scientist and mathematician James S. Williamson was instrumental in developing one of the first solar-powered electricity-generating plants in the United States. Throughout his career Williamson, a member of the Chippewa nation, has taught at the university level and has helped formulate new methods for educating students in mathematics. He has also been active in government, having been appointed by President Jimmy Carter to the Domestic Policy Review Panel for Solar Energy Research.

Williamson, the eighth of twelve children, was born on November 30, 1949, in Williston, North Dakota. His mother was a homemaker and his father was a civil servant. In college he studied mathematics, earning a bachelor of science degree from Montana State University in 1971 and a master of science degree from the University of California at Berkeley in 1974. His research at Berkeley involved hyperbolic (relating to the hyperbola, a plane curve) and elliptical (relating to the ellipsis, a closed curve) functions of imaginary numbers, an area of research that could be applied only to space travel at the time. Williamson also taught undergraduate courses in mathematics and served on a team that developed new methods for teaching mathematics.

IMPACT

James S. Williamson, who developed one of the first solar-powered electricity-generating plants in the United States, is actively involved in developing science, mathematics, and technology education for students of all ages, from kindergarten through college, by using solar energy programs. He designs curriculum models for school systems and organizes and manages student competitions. One of his projects was a mini solar-powered car competition, complete with ramp tests, speed tests, and a design contest.

Begins solar energy research

Williamson began his career in solar energy, or heat energy provided by the Sun, research in 1971 as a project manager for the Atomic Energy Commission, first in Idaho Falls, Idaho, then in Oakland, California. Originally he had been hired as a mathematician and statistician to evaluate how to clean up fuels released from a spent nuclear reactor, using mathematical models. When an opportunity arose to work on space projects he took it, a move that brought him into the field of solar energy. Among the projects he managed for six years were power systems for satellites and a nuclear reactor system. He also directed the design of the central receiver test facility in Albuquerque, New Mexico.

Williamson then supervised the design of the first U.S. solar central receiver electricity-generating plant in Barstow, California. The Barstow plant has more than one thousand large mirrors, each of which reflects onto a central steam boiler mounted on a tower several hundred feet in the air to make steam. The steam, in turn, drives a turbine (rotary engine) to generate electricity for the Los Angeles area.

Moves to private industry

In 1977 Williamson moved to the Martin Marietta Corporation in Denver, Colorado, as deputy manager for advanced solar technologies. After a year he joined Midwest Research Institute as program director for international solar energy programs. He held that position for eight years, spending the first four years in Golden, Colorado, and the next four in Kansas City, Missouri. His research concentrated on water desalination (removal of salt), independent utility power systems, and solar systems. Throughout his career in government and private industry he never abandoned teaching. During his last two years at Midwest he taught evening courses in mathematics at the University of Missouri in Kansas City.

After leaving Midwest in 1986 Williamson became an associate partner and director at the Meridian Corporation in Washington, D.C., where he headed projects for the Department of Energy. He also conducted biomass (plant material and animal waste used as fuel) resource surveys and developed a technology assessment guide for public utilities. Williamson served briefly as president of Engineering and Product Services, a small engineering consulting firm, where he designed a recruiting strategy for Native American health centers.

Returns to education

The late 1980s marked another shift in Williamson's career. He became involved in using solar energy programs and activities to increase young students' interest in science and mathematics. Since 1989 he has been assistant to the director for education and special projects at the National Renewable Energy Laboratory. Initially based in Washington, D.C., he returned to Golden, Colorado, in 1992. He has been an adjunct lecturer on the business and economics of alternative energy technologies at Regis University in Denver, Colorado, and has lectured on thermodynamic applications in solar energy at the Air Force Academy in Colorado Springs, Colorado. His most recent work has been the development of science, mathematics, and technology education for students from kindergarten to the university level.

Honored for work in solar research

Williamson's other achievements include directing joint solar energy research projects between the United States and

Maria Telkes, Hungarian-born American Physical Chemist

Hungarian-born American physical chemist Maria Telkes (1900–) devoted most of her career to investigating solar energy and designing devices to take advantage of the heat energy provided by the Sun. Telkes first became interested in solar energy in 1939, when she joined the Massachusetts Institute of Technology (MIT) Solar Energy Conversion Project. There she researched and designed a new type of solar heating system that stored solar energy as chemical energy through the crystallization of a sodium sulphate solution rather than by heating water or rocks, as was previously done. Her innovative heating system was installed in a prototype house built in Dover, Massachusetts, in 1948.

Telkes was soon recruited by the U.S. government to develop ways to produce drinking water from seawater. In the desalination process, by which salt is removed from seawater, the water must be vaporized into steam, and then the steam must be condensed (changed into liquid) to give pure water. Telkes designed a still that utilized solar energy for vaporizing the water that could be installed on life rafts. This design was enlarged for use in the Virgin Islands, where the fresh water supply is often endangered.

Telkes eventually returned to private industry, where she developed solar dryers, solar water heaters, and solar thermoelectric generators in space. In 1969 she joined the Institute of Energy Conversion at the University of Delaware, where she developed materials for storing solar energy and designed heat exchangers for efficient transfer of energy. Her advancements, which resulted in a number of domestic and foreign patents for the storage of solar heat, were put into practical use in Solar One, an experimental solar-heated building at the University of Delaware.

Saudi Arabia, conducting research on air conditioning systems to generate electricity from sunlight, and developing material safeguards systems for nuclear power plants. In 1983 he received the Management Award from the Saudi Arabian National Center for Science and Technology for his work on this joint project. He has also designed nuclear and solar power systems for space travel to Saturn, Mars, and Jupiter.

Williamson is a charter member of the American Indian Science and Engineering Society, as well as a member of the

American Solar Energy Society. He holds memberships in the Institute of Electrical and Electronic Engineers and the American Management Association. In 1991 he was presented the Organization Committee Award at the Conference for Economics and Ethanol Fuels at Montana State University. Williamson married Virginia Hansen in 1968. They have two children, James and Kerry.

Further Reading

American Indian Science and Engineering Society Biography Booklet, 1984–1985, p. 29.

Directory of the National Society of the Sons of the American Revolution, 1993.

Edward O. Wilson

Born June 10, 1929
Birmingham, Alabama

World-renowned entomologist Edward O. Wilson, nicknamed "Dr. Ant" for his study of insects, is a founder of the field of sociobiology. Sociobiologists study the way animal societies function and try to determine how such societies have come about. While his theories are considered controversial, he has made significant contributions to research on animal and human social behavior. Wilson has also influenced the field of animal taxonomy (classification) through his observations of insect species and has conducted studies that led to the discovery of pheromones, chemical substances produced by animals that are instrumental in communication among individuals of the same species. Wilson has also become active in the environmental movement and is largely responsible for bringing the issue a biodiversity to a wide audience through his 1992 book *The Diversity of Life*. Wilson is the recipient of two Pulitzer Prizes for literature, the 1977 National Medal of Science, and the 1990 Crafoord Prize, which is equal in stature to the Nobel Prize.

"My role was to introduce evolutionary biology in human development; now the ball is rolling and it's not my field. The social insects are my true love."

IMPACT

With the publication of *The Diversity of Life* in 1992, American entomologist and sociobiology pioneer Edward O. Wilson brought the issue of biodiversity for the first time to a mainstream audience. The book, which calls for increased protection of the environment to minimize the mass extinction of species, is considered to be as revolutionary as **Rachel Carson**'s *Silent Spring* (see entry), which thirty years earlier warned about the ecological hazards of toxic chemicals. Scientists, environmentalists, and policymakers alike regard *The Diversity of Life* as an important step that will focus international attention on biodiversity into the next century.

Decides to study insects

Edward Osborne Wilson was born on June 10, 1929, in Birmingham, Alabama. A descendant of farmers and shipowners, Wilson had decided by the age of seven that he wanted to be a naturalist and explorer. Fate intervened, however, during a fishing trip. As he was vigorously pulling a fish out of the water, its fin grazed and damaged his right eye. Thereafter he developed the habit of closely examining animals and objects with his keen left eye. When he was ten he read a *National Geographic* article titled "Stalking Ants, Savage and Civilized," and became interested in entomology, the branch of zoology that deals with insects. Wilson studied biology at the University of Alabama, obtaining a bachelor of science degree in 1949 and a master of science degree in 1950. Five years later he received his doctorate in biology from Harvard University in Cambridge, Massachusetts. Wilson gained full professorship in 1964 and was named Frank B. Baird Jr. Professor of Science in 1976.

Introduces new scientific theories

During the early years of his career Wilson worked in the field of systematics, the attempt to classify species based on the principles of evolutionary theory. With his Harvard colleague William L. Brown he introduced new procedures for taxonomists, or scientists who classify organisms into specially named groups based on shared characteristics and natural relationships. In 1956 Wilson also codeveloped the concept of "character displacement," which occurs when two similar species begin a process of genetic differentiation (development of new genetic characteristics) to avoid competition and cross-breeding.

In the middle to late 1950s Wilson traveled to Australia, the South Pacific islands, and Melanesia to study and classify ants native to those regions. As a result of his fieldwork in the Melanesian archipelagoes (islands scattered in the ocean), he developed the concept of the taxon cycle. Using Melanesian ants as his subject, Wilson described the taxon cycle as the process by which new species are formed. As a species moves to a new, harsher habitat, it evolves into one or more new "daughter" species, which then adapt to the new habitat. The taxon cycle has since been found among other insects and birds.

Discovers pheromones

In 1959, while studying a colony of fire ants, Wilson discovered that social insects such as ants communicate through chemical releasers. Influenced by the rise of molecular biology, Wilson set out to prove his hypothesis by crushing a venom gland he had extracted from a fire ant and creating a trail of the chemicals from the gland near a colony of the same species. He had anticipated that only a few ants would trace the chemical path, but instead dozens of fire ants swarmed out of the colony and followed the trail. After close observation Wilson discovered that the chemicals communicate complex instructions from one ant to another—everything from the location of food and how to obtain it to a call for help when in distress. The chemicals came to be known as pheromones, and Wilson's discovery launched extensive research on the behavior of social insects that continues today.

Identifies species equilibrium

In the early and mid-1960s, collaborating with Princeton University mathematician Robert H. MacArthur, Wilson developed the first mathematical theory of species equilibrium. He and MacArthur hypothesized that the number of species on a small island remains constant, even though the variety of species undergoes constant reshuffling. Scientists had already established that two factors, extinction and immigration, affect the number of species in an ecosystem (a community of organisms and its environment that functions as a unit in nature). Applying this concept to their study of islands, Wilson and MacArthur suggested that extinction (the total disappearance of a species in a given area) and immigration (movement by a species to a new habitat) are determined by the size and closeness of the islands. For example, larger, less-crowded islands typically have lower extinction rates, and islands that are close together experience greater species immigration from one island to another.

Calling this phenomenon "equilibrium hypothesis of island biogeography," Wilson and MacArthur formulated a mathematical model to describe the relationship between extinction and immigration. They depicted the rate of extinction with a positive (upward) sloping curve and the rate of immigration with a negative (downward) sloping curve. They predicted that the actual number of species could be found at the intersection of the two curves. In 1968 Wilson confirmed this hypothesis with biologist Daniel Simberloff in a study of insect life on six islands off the Florida Keys. They first counted the number of insect species on the islands, then fumigated (used a chemical spray to kill insects) the islands and recounted the species eight months later. As Wilson had predicted in the mathematical formula, the number of species remained the same while the composition of individual species was significantly different, and did in fact evolve over time. For this landmark work Wilson received the 1990 Crafoord Prize, awarded by the Royal Swedish Academy of Sciences.

Founds field of sociobiology

But Wilson's crowning achievement was his 1975 book *Sociobiology: The New Synthesis,* in which he defines the science of sociobiology as "the systematic study of the biological basis of all social behavior." The term had been in use prior to Wilson's book, but he gave it a new dimension. Many scientists consider his concept to be as revolutionary as the theory of natural selection developed by the English naturalist **Charles Darwin** and the space-time theory formulated by the

Simon LeVay, American Neurobiologist

Research performed by American neurobiologist Simon LeVay (1943–) appears to support Edward O. Wilson's hypothesis that social behavior may be genetically based. Working at the Salk Institute in San Diego, California, LeVay believes that homosexuality (attraction to members of the same sex) could be biologically determined. The researcher bases his theory on his studies of forty-one male and female brains. He examined the area of the hypothalamus, located in the center of the brain, in the area that is known to help regulate male sexual behavior. Within this site LeVay looked at four different groupings of cells, technically referred to as the interstitial nuclei of the anterior hypothalamus, or INAH for short. Other researchers had already reported that INAH 2 and INAH 3 were larger in men than in women. LeVay hypothesized that one or both of the nuclei might vary with sexual orientation as well.

Nineteen of the brains LeVay examined had come from homosexual men who died of AIDS. After careful examination of the brain samples, he found that the INAH 3 areas of most of the women and homosexual men were about the same size. In heterosexual men this region was, on the average, twice as large—or about the size of a grain of sand. For LeVay, who is himself homosexual, the genetic findings may help turn around a longstanding stereotype about the "causes" of homosexuality. Possible contradictions of the scientist's findings, however, include the idea that AIDS itself had caused the discrepancy in size of the INAH 3 nuclei in the brains LeVay studied. Several of LeVay's colleagues nonetheless support his views. Laura Allen, a scientist of the University of California at Los Angeles, said LeVay is "probably correct" in his theory, since the heterosexual men's brains contained larger INAH 3 nuclei whether or not they had died of AIDS. LeVay and others hope their work will enable humans to view homosexuality the way other species see it: as a normal variation of sexual behavior.

German-born American physicist **Albert Einstein** (see entries). (Darwin's theory of natural selection is the principle that only the organisms best suited to their environment tend to survive and pass on their genetic characteristics to their offspring. Einstein's space-time continuum posits that space is curved, and that a body affects the shape of the space that surrounds it so that a second body moving near the first body will travel in a curved path.) In fact, many of the tenets Wilson put forth in his book have gained widespread acceptance—while some have aroused considerable controversy.

Wilson's journey into sociobiology began early in his career, when he was in the midst of an exciting development, called the new synthesis, that reinvigorated evolutionary biology. Following World War II, in the 1940s and 1950s, breakthrough advances in the sciences allowed new genetic data to be applied to biological studies that previously were highly descriptive and often unsystematic. At that time the field of molecular biology, or the study of the structure and function of molecules that make up living organisms, was also gaining prominence. But Wilson observed that the rise of molecular biology threatened to relegate the softer study of animal behavior to a tiny corner of Harvard's biology department. Thus he began focusing on the significance of organisms as carriers of genetic information. Viewing the complex behavior of ants and other social insects in this framework prompted Wilson to describe behavior that served survival not of the individual, but of the population.

Eventually, Wilson was able to explain such social characteristics as altruism (concern for the welfare of others), significance of kinship, communication, and specialization of labor, which had previously puzzled scientists. According to Wilson, cooperation among individuals or between species is consistent with early evolutionary theory because it enables individuals to survive and carry on the gene pool (the collection of genes of all the individuals in an interbreeding population). But up to that time altruism and spiteful behavior (when an individual harms another and itself) were largely unexplained by biologists.

Explains altruism in ants

In order to account for altruism and spiteful behavior, Wilson focused on population survival among ants. He found that in a colony of ants, sterile members will work for their family members, who share similar genes and will reproduce on their behalf. Wilson maintains that selflessness is a characteristic of most ant species. He describes their colonies as "superorganisms" in which the welfare of the colony, not the individual, is of greatest importance. Similarly, but on the other end of the behavioral spectrum, a species of Malaysian ants will rupture glands of poison on their own bodies if invaded by enemies, killing themselves and their intruders while signaling for help from members of their own colony.

Other complex and intricate behaviors are explained by Wilson's sociobiology as well. The European red Amazon ant, for example, is an aggressive creature that actually invades the nests of more peaceful ant species, killing some individuals and capturing others for use as slaves in its own nests. The slave ants actually do "housework," digging chambers and feeding and nurturing the young Amazons. Thus the Amazon ants are able to ensure the survival of their own populations through aggression and dominance toward ants outside their colony.

Applies sociobiology to humans

It was the twenty-seventh chapter of *Sociobiology,* titled "Man From Sociobiology to Biology," that touched off a controversy that continues today. Advocating expanded research on the role of biology in human behavior, Wilson wrote, "There is a need for a discipline of anthropological genetics. By comparing man with other primate species, it might be possible to identify basic primate traits that lie beneath the surface and help to determine the configuration of man's higher social behavior." Wilson noted that humans have always been characterized by "aggressive dominance systems, with males generally dominant over females." He also commented that "a key early step in human social evolution was the use of women in barter." In a separate article Wilson wrote: "In hunter-gatherer societies, men hunt and women stay at home. This strong bias ... appears to have a genetic origin. Even with identical education and equal access to all professions, men are likely to continue to play a disproportionate role in political life, business, and science."

Sociobiology, and especially chapter twenty-seven, sparked a national debate. Wilson witnessed the extent of the controversy at a 1978 meeting of the American Association for the Advancement of Science, where he was scheduled to speak. Moments before his presentation, hecklers took the stage and doused him with a pitcher of water, shouting, "We think you're all wet!" At that meeting several of his colleagues signed a letter of protest. They asserted that in the past, theories such as his were responsible for the "sterilization laws and restrictive immigration laws by the United States between 1910 and 1930 and also for the eugenics policies which led to the establishment of gas chambers in Nazi Germany." (Eugenics is a science that deals with the improvement of hereditary qualities of a race or breed by such practices as control over mating or sterilization. Adolf Hitler and his Nazi Party, who came to power in Germany prior to World War II, espoused the superiority of the Aryan, or white, race. During the war the Nazis killed more than six million Jews, whom they considered of inferior genetic stock, by sending many of them to gas chambers.)

Wilson's harshest critics believed that the inevitable conclusion of his theories was "biological determinism." **Stephen Jay Gould** (see entry), a fellow Harvard professor who had signed the letter, sought a middle ground. He pointed out that Wilson's theory could also be interpreted as showing that "peacefulness, equality, and kindness are just as biological as violence, sexism and general nastiness." Gould maintained, however, that there is no direct evidence to prove that specific human behaviors are genetically determined.

Wins Pulitzer Prizes

In 1978 Wilson published *On Human Nature,* in which he clarified the controversial views he had expressed in *Sociobiology* and gave a fuller explanation of such issues as free will, ethics, and human development. The book won the 1979 Pulitzer Prize for general nonfiction. He later collaborated with University of Toronto professor Charles Lumsden in writing *Genes, Mind and Culture* (1981) and *Promethean Fire* (1983), which Wilson describes as his "last word on the subject" of human sociobiology. His life-long fascination with ants culminated with the publication of 1990's *The Ants,* which he wrote with German entomologist Bert Holldobler. This book also earned Wilson a Pulitzer Prize for general nonfiction.

Warns of ecological destruction

In recent years Wilson has turned his attention to the environment and ecology, the branch of biology that studies the relationships between living things and their environments. In 1992 he published *The Diversity of Life,* which brought the issue of biodiversity to a mainstream audience for the first time. Biodiversity is reflected by the numbers of different species of plants and animals in an environment. In his book Wilson argues for increased protection of the environment to minimize the mass extinction of species already underway. The actual number of existing species and the exact rate of species decline is a topic of heated debate in the scientific community, but researchers calculate there are more than 10 million and possibly even 100 million species of plants and animals on Earth, yet only 10 million have been identified. Wilson estimates that 1 out of every 1,000 species becomes extinct each year. He also predicts that in the rain forest alone, one-half the species of plants and animals could be extinct by the mid-twenty-first century, a rate that will rival the extinction that marked the end of the dinosaur age. Wilson advocates taking an inventory of plant and animal life, then promoting the wise use of plant and animal resources for food and medicine. He also urges controlled development and restoration of land that has already been damaged.

Continues writing

In addition to writing books, Wilson has contributed more than three hundred articles to publications as diverse as *National Geographic, Omni,* the *New York Times, American Zoologist,* and *BioScience.* His topics range from shark adaptations and the rain forest canopy to the decline of bird populations. Although Wilson is often on the road speaking about biodiversity to academics, environmentalists, and political leaders, he has not lost his curiosity about the natural world. In *The Diversity of Life* he described an encounter he had in a rain forest one night: "I swept the ground with the beam from my headlamp for signs of life, and found—diamonds! At regular intervals of several meters, intense pinpoints of white light winked on and off with each turning of the lamp. They were reflections from the eyes of wolf spiders, members of the family Lycosidea, on the prowl for insect prey." Wilson has also managed to find time to write his autobiography, *Naturalist,* which was published in 1995.

Honored for work in sociobiology and ecology

Included among Wilson's numerous honors and awards are the Cleveland Prize (1967), the Mercer Award of the Ecological Society of America (1971), the Founders' Memorial Award from the Entomological Society of America (1972), the Leidy Medal (1978), the Carr Medal (1978), the L. O. Howard Award of the Entomological Society of America (1985), and the Tyler Prize for Environmental Achievement (1984). Wilson served on the World Wildlife Fund Board of Directors from 1984 to 1990. He is a member of the National Academy of Sciences, a fellow of the American Academy of Arts and Sciences and of the American Philosophical Society, the former president of the Society for the Study of Evolution, and an honorary member of the British Ecological Society.

Further Reading

American Zoologist, February 1994.

BioScience, December 1995.

Boys' Life, February 1993.

Discover, March 1990; November 1993.

National Geographic, December 1991.

Nature Conservancy Magazine, July/August 1994.

New York Times, May 13, 1993.

New York Times Book Review, October 4, 1992.

New York Times Magazine, July 22, 1990; May 30, 1993.

Omni, September 1990.

Science, September 1, 1995.

Scientific American, April 1994.

Wilson, Edward O., *The Diversity of Life,* Belknap Press/Norton, 1992.

Wilson, Edward O., *Sociobiology: The New Synthesis,* Belknap Press, 1975.

Granville T. Woods

Born April 23, 1856
Columbus, Ohio

Died January 30, 1910
New York, New York

Granville T. Woods "has left us the rich legacy of a life successfully devoted to the cause of progress."

—Henry E. Baker

Granville T. Woods was one of the most prolific and visionary inventors of the American Industrial Revolution. At the time of his death in 1910, Woods had been granted sixty patents, many of which revolutionized railway and telegraph communication—and made his competitors, such as Alexander Graham Bell, Thomas Alva Edison, and George Westinghouse, rich. A black man who faced the prejudice of post-Civil War America, Woods could never generate the capital necessary to manufacture his own inventions, so he had to sell his patents to larger corporations. Consequently, he never received the credit he deserved during his lifetime, and he died nearly penniless.

Travels the world

Granville T. Woods was born on April 23, 1856, in Columbus, Ohio, five years before the start of the Civil War. He was born free under the Northwest Ordinance of 1787, which outlawed slavery from the territory that included the future state of Ohio. However, in 1803 Ohio adopted "Black Codes" that restricted the participation of blacks in the state militia, in certain legal matters, and in public education. By the time Woods began attending school in the 1860s the state had modified its ban on public education, but the lives of blacks were still severely regulated. At the age of ten Woods was forced to leave school and apprentice as a machinist and blacksmith in a machine shop. This was where he began his lifelong interest in electrical and mechanical engineering.

In 1872, at age sixteen, Woods left Ohio and worked various jobs around the United States. First he worked as a fireman and as an engineer for the Iron Mountain Railroad in Missouri, where he became interested in electricity and its application to railroads. In 1874 Woods moved to Springfield, Illinois, to take a job in a rolling mill, a factory where metal is rolled into sheets. Then to increase his working knowledge of mechanics and electricity, Woods qualified to enroll in courses in mechanical and electrical engineering at an eastern college in 1876. He attended classes at night while he worked during the day at a New York City machine shop. After two years he left school and signed on as an engineer aboard the British steamer *Ironsides,* which, over a two-year journey, took him to nearly every continent. In 1880 he returned to the United States to work as a steam locomotive engineer for the Danville and Southern Railroad in Cincinnati, Ohio, a position he held for four years.

Although his work record and his education should have entitled him to more responsible positions, Woods was constantly denied them. Even though the Civil War, which had freed blacks from slavery, had been over for nearly twenty years, the attitude that blacks were inferior to whites had hardly changed. Fortunately, Woods's knowledge of mechanical and electrical applications helped him. In 1884 he received his first patent, for a more efficient version of a steam boiler furnace. That same year, with his brother Lyates, Woods opened the Woods Electric Company in Cincinnati to produce and market his own inventions.

Inventor limited by patent laws

Even though Woods received patents for many of his inventions, the legalities of patent assignment and regulation did not protect his work. The U.S. government grants patents for any new or useful machine, process, or manufacturing method, or for an improvement of any previous machine or process or method. A patent gives its owner, the patentee, the sole right to manufacture, use, and sell that particular invention. Patents are, in a sense, recognized as personal property. If others try to make, use, or sell a patentee's invention, they are guilty of patent infringement and may be sued by the patentee. There are two instances, however, when an inventor may lose his right to a patent for his invention. First, if an inventor does not have enough money to manufacture and market his invention, the patent is then assigned or sold to another individual or company that has the necessary capital. Second, the inventor may simply wish to sell his patent outright. In either case, once the patent is assigned to someone else, the inventor gives up all legal and monetary claims to that invention. Because Woods was black and was unable to secure a high-paying job, he couldn't raise the capital to manufacture and market his inventions himself. As a result, Woods would have to sell his patents to large companies and was not able to earn profits from the products.

Develops first two inventions

Woods's first two electrical inventions dealt with sound transmission. In December 1884 he was granted a patent for a telephone transmitter, an apparatus that conducted sound over an electrical current. **Alexander Graham Bell** (see entry) had already developed a telephonic device almost a decade earlier, but Woods's instrument far surpassed any models then in use because it carried a louder and more distinct sound over a longer distance. The physical properties by which the device operated are still employed in modern telephones. Despite this great achievement, patent guidelines dictated that the patent be assigned to a company that had the mechanical and monetary means to manufacture such a device. Consequently, the patent was not assigned to Woods, but to the American Bell Telephone Company.

Less than a year later Woods was granted a patent for a mechanism he called a telegraphony. This device was a combination telegraph and telephone that could transmit both oral and signal messages. Prior to this invention the telegraph could send messages only over an electrical current utilizing Morse code (see box), a combination of short and long pulses commonly referred to as dots and dashes that represent letters of the alphabet, which became the language of the telegraph. Thus, in order to send and receive messages, the operators on either end of a telegraphic transmission had to be fully versed in Morse code and in the operation of the sending key apparatus.

Woods telegraphony, however, gave almost everyone, regardless of their knowledge of telegraphs, the chance to send messages. Anyone unfamiliar with Morse code could simply flip a switch on the telegraph and speak into a mouthpiece near the sending key. The message would then be heard on the receiving end as articulate speech. Because of the great demand for his invention, Woods again sold his patent to the American Bell Telephone Company, which paid him generously.

Over the next twenty-five years Woods's inventions were numerous and varied, from an incubator that provided a constant temperature for the hatching of chicks to a series of tracks used by motor vehicles at amusement parks. He also invented a grooved-wheel system for transferring electric current to streetcars; the wheel, called a troller, is the source of the popular name of the streetcar, the trolley.

IMPACT

Although he was one of the most prolific inventors of the American Industrial Revolution, Granville T. Woods was never able to raise the capital to manufacture his own inventions. He was forced to sell many of his sixty patents—many involving railway and telegraph communication—to larger corporations. He thus contributed in a large way to the success of such inventors and industrialists as Thomas Alva Edison and Alexander Graham Bell. Woods's greatest invention was the Induction Telegraph System, an improved system that allowed trains—sometimes heading toward each other on the same track—to communicate with one another. He also invented the telegraphony, a combination telegraph and telephone that gave people unfamiliar with Morse code the opportunity to send messages across telegraph lines. Despite his inventiveness, Woods never received the opportunities or the recognition afforded his white counterparts in post-Civil War America.

Samuel F. B. Morse, American Artist and Inventor

American artist and inventor Samuel F. B. Morse (1791–1872) perfected the electric telegraph and the telegraph code named for him. Morse was a well-known portrait painter who, in 1837, turned his attention to science to realize his ideas for the telegraph, which he had first learned about on a ocean voyage in 1832. He had built several prototypes of the device by 1835, but his lack of technical background hampered his efforts. While he was on the arts faculty at the University of the City of New York, he turned to chemistry professor Leonard Gale for help. In addition to giving him practical advice, Gale introduced Morse to Princeton University physicist Joseph Henry, who freely shared with Morse his knowledge of electromagnetism.

The telegraph uses a battery (a device that furnishes electric current) and an electromagnet (a core of magnetic material surrounded by a coil of wire through which an electric current is passed to magnetize the core) that raps against a metal contact when activated. Gale showed Morse how to supply sufficient current in both his battery and electromagnet. Morse was now able to design a relay system (a system of automatic controls) using a series of electromagnets to open and close circuits along the telegraph wire, which kept the current strong enough to travel long distances.

In 1837, while demonstrating the telegraph in New York, Morse met a young inventor named Alfred Vail, and the two men became partners. Vail made many practical improvements to Morse's telegraph and helped refine the message-transmitting system that became known as the Morse code. Morse code is really a series of short electrical impulses—read as dots and dashes that represent letters, numbers, and punctuation—that repeatedly activate and interrupt this magnetism, resulting in a tapped-out message. In 1837 Morse applied for a grant offered by the U.S. Congress to construct a telegraph system; seven years later, in 1843, Morse finally secured the $30,000 to build a telegraph line between Baltimore, Maryland, and Washington, D.C. He sent his first message—"What hath God wrought!"—on May 24, 1844. Over the next decade Morse would become embroiled in lawsuits over patent rights to the telegraph. Although there was great controversy and hostility—Morse even denied that Henry had helped him with the original design—Morse's patent rights were upheld by the U.S. Supreme Court in 1954. The telegraph would become the standard way of communicating within cities until the telephone became widely available in the 1880s.

Saves many lives

Woods's greatest invention improved electrical communication between trains. On November 29, 1887, Woods received a patent for his Induction Telegraph System, also called the Synchronous Multiplex Railway Telegraph. Before Woods's invention communication between moving trains and between a moving train and a railroad station had been poor. In a telegraph system a continuous wire must exist between a sending key and a receiving sounder. Ordinary telegraph wires were usually run along railroad tracks, but for a telegraph system to work aboard the train, part of the train had to have been in constant contact with these wires. Because of the jostling movement of trains, most messages sent or received were incomplete. Numerous times, warnings of washed-out bridges, rock slides, and other obstructions failed to reach a train in time. Other times, trains learned too late—or not at all—of the location of other trains on the same track.

Wood's Induction Telegraph changed the course of railway travel and saved many lives. To realize his invention, Woods applied the law of electromagnetic induction formulated by **Michael Faraday** (see entry). The device consisted of an oblong coil suspended beneath a train, through which an electrical current was passed. In turn, a magnetic field developed around the train. When the train moved, the field moved with it and induced a similar current in the telegraph wires that ran along the tracks. This mechanism allowed uninterrupted sending and receiving of telegraphic messages on moving trains.

Faces controversy

Woods received considerable recognition for this invention. However, controversy developed when **Thomas Alva Edison** (see entry) and another inventor, Lucius Phelps, claimed that they each had developed a similar telegraph system before Woods. In both cases Woods was declared the prior inventor. Nonetheless, he faced more legal challenges in 1892 when he was sued for criminal libel after he claimed that a manager of the American Engineering Company stole his patent for an electric railway. He was jailed briefly when he could not post money for bail.

Woods's payment of large legal fees, along with the loss of income from his inventions, left him in poverty at the end of his life. As the owner of a small company, he could hardly compete with the larger corporations like those of Edison and Bell. Also, as a black inventor Woods could not hope to receive the deserved public recognition his white counterparts enjoyed. All this, however, did little to diminish his intellectual and creative output. Woods died in New York City in 1910.

Further Reading

Baker, Henry E., *The Colored Inventor,* Crisis Publishing Company, 1913, reprinted, Arno Press, 1969.

Current, Richard N., T. Harry Williams, Frank Freidel, and Alan Brinkley, *American History, A Survey–Volume 1: To 1877,* 6th edition, Knopf, 1983.

Haber, Louis, *Black Pioneers of Science and Invention,* Harcourt, 1970.

Haskins, Jim, *Outward Dreams: Black Inventors and Their Inventions,* Bantam, 1992.

Hayden, Robert C., *Eight Black American Inventors,* Addison-Wesley, 1972.

Jackson, W. Sherman, "Granville T. Woods: Railway Communications Wizard, 1856–1910," in *American Black Scientists and Inventors,* edited by Edward S. Jenkins, National Science Teachers Association, 1975.

James, Portia P., *The Real McCoy: African-American Invention and Innovation, 1619–1930,* Smithsonian Institution Press, 1989.

Wilbur Wright

Born April 16, 1867
Millville, Indiana
Died May 30, 1912
Dayton, Ohio

Orville Wright

Born August 19, 1871
Dayton, Ohio
Died January 30, 1948
Dayton, Ohio

American inventors Wilbur and Orville Wright were not the only ones to attempt powered flight, but they were the first to conduct serious wind tunnel experiments. These experiments enabled them to predict how a plane would fly through the air. Because of their methodical observations and sharp mechanical skills—not to mention a relentless determination—they succeeded in building the first working airplane.

"From the time we were little children my brother Orville and myself lived together, played together, worked together, and in fact thought together."

—Wilbur Wright

Build their first flying machine

Wilbur Wright was born on April 16, 1867, near Millville, Indiana, and his younger brother, Orville, was born on August 19, 1871, in Dayton, Ohio. Their father, Milton Wright, was a bishop in the United Brethren in Christ Church and later became a theology professor and editor of a weekly religious newspaper in Dayton. The boys' mechanical aptitude came from their mother, Susan. Being inventive and good with tools, she once single-handedly built a sled for the children.

IMPACT

Wilbur and Orville Wright were American inventors with little formal training who are best remembered for inventing the first heavier-than-air, engine-powered passenger flying craft. They not only invented the airplane but also developed some of the systems we use for flight control today. Had it not been for a bit of over-control of his craft, Wilbur might have been the one who actually flew first, instead of Orville. The brothers were constantly curious and persistent tinkerers, two qualities that were vital in overcoming the obstacles to flight posed at the time of their endeavors.

Milton Wright also provided inspiration. When Wilbur and Orville were young, he came home with a surprise for them: a toy helicopter powered by a rubber band. The boys were amazed with the bamboo-and-paper toy, which had been created by the French inventor Alphonse Penaud. They tried to make a larger version of the toy, thinking that it would fly higher. Instead, they discovered that the opposite was true: the larger their helicopter model, the more trouble it had flying. This surprised the boys considerably and led to some of their earliest musings on flight.

When Orville was fourteen, he and a friend set up a printing shop in his friend's basement and published a school newspaper called *Midget.* Two years later Orville built his own printing press, and after Wilbur made several improvements to the design, the two began printing the church newspaper. Using another home-built press, the Wright brothers published the *West Side News,* with Orville as business manager and Wilbur as editor. The *West Side News* began as a weekly but eventually evolved into a daily paper called *The Evening Item.* When that venture failed, the Wrights established a business printing religious pamphlets. Among their clients was Paul Lawrence Dunbar, a friend of Orville and publisher of *The Tatler,* a newspaper for the black community of Dayton. Dunbar later became a famous poet.

Design and manufacture bicycles

While they were in their early twenties, Wilbur and Orville ran a bicycle repair shop in Dayton and eventually began designing and manufacturing their own models. The business was profitable and kept them busy, but their inventive minds drove them to work on other projects. Besides making

How an Airplane Flies

An airplane obtains lift from the aerodynamic effect of the air rushing over its wings. (Aerodynamics concerns the flow of air or other gases around a body in motion.) Lift occurs because the wing's upper surface is more convex (curved downward), and therefore longer, than the lower surface of the wing. As a result, air must travel faster past the upper surface than past the lower, which leads to reduced pressure above the wing. Since the pressure below the wing is stronger than that above, the plane is pushed aloft. The lift provided by the wings must equal the aircraft's weight; the forward thrust of the engine must balance the forces of drag on the plane that result from its movement through the air. In modern flight, there are many other factors to consider as well, but most important is that all the forces act on an airplane's balance.

continual improvements to the bicycle, they also made refinements to adding machines and typewriters. Orville even toyed with the idea of starting an automobile manufacturing business, but Wilbur thought it would be easier to build a flying machine.

Begin to research flight

The Wright brothers had heard of glider experiments on both sides of the Atlantic. Otto Lilienthal (see box) in Germany and Octave Chanute in the United States, both considered pioneers in aviation, were also heroes in the eyes of the Wrights. In 1896 Orville came down with typhoid fever. During his long recovery he and Wilbur discussed the possibility of building their own glider, an aircraft similar to an airplane but without an engine. The prospect excited the brothers, and they read everything they could find on the subject. They even wrote the Smithsonian Institution for additional information on gliders. The Smithsonian, a U.S. government-sponsored center for scientific and artistic culture, responded with a list of current books on flight, including Chanute's *Progress in Flying Machines* and *Experiments in Aerodynamics* by American astronomer and physicist Samuel P. Langley (see box).

Otto Lilienthal, German Aeronautical Engineer

German aeronautical engineer Otto Lilienthal (1848–1896) was a major figure in the ten years preceding powered flight. His experiments with gliders were followed all over the world by the press and aviation enthusiasts including the Wright brothers. As a young man Lilienthal and his brother studied birds, particularly storks, to unlock the secret of flight. They even built strap-on wings in the hopes of flying, but failed in their attempts.

Lilienthal came to realize that a curved wing was necessary to succeed in flight. Measuring the amount of lift that various rigid wings produced, Lilienthal compiled tables of calculations and published findings in an 1889 book that became indespensable for other aviation enthusiasts: *Bird-flight as the Basis of Aviation.* Lilienthal then applied what he knew about gliding to powered flight. Starting with short jumps, he soon needed longer and higher areas to work. In 1894 he built an artificial hill, where he achieved glides of 150 feet or more. He also found hills outside of Berlin, where he reached a gliding distance of 1,150 feet.

Despite these successes, Lilienthal probably would never have achieved successful powered flight because he maintained that the only way to move through the air was by flapping wings. He devoted his energies to building machines with flapping action, but they were so ineffective that he was never able to try them. In August 1896 tragedy struck when a sudden gust of wind caused Lilienthal's glider to rise suddenly. Although he threw his weight forward to correct the problem, the glider stalled and crashed to the ground, and he died the next day.

Use birds as example

One of the central problems with glider flight was balance. Glider pioneers were able to keep their wings balanced by shifting their weight to the left or right as needed. This was a tiring activity, however, requiring near-acrobatic skill. When he was watching buzzards, Wilbur noticed that birds keep their balance during flight by changing the curvature of their wings. From this observation he concluded that a working airplane would have to do the same with its wings—that somehow the wings should be able to alter their shape in flight. One day, while fiddling with an empty cardboard box at the bicycle shop, Wilbur realized how the wings could be designed.

During the summer of 1899 the Wright brothers built a large kite in the form of a biplane, an aircraft having two sets of wings, one above the other, which would be rigid near the center of the plane but able to move in opposing angles at the tips. By warping the wings, they were able to make the kite climb or dive. Once they could successfully control the kite, the brothers decided to build a glider, and they began looking for a site to conduct test flights.

Test early biplane

In December 1899 Orville and Wilbur wrote to the U.S. Weather Bureau for information about wind velocity (speed) in different geographic locations. Specifically, they were seeking a place with good weather, fifteen-mile-per-hour winds, and a sandy, treeless terrain. The Weather Bureau recommended Kitty Hawk, a small fishing town on a strip of sand,just off the mainland of North Carolina. In September of the following year the Wright brothers made their first six-hundred-mile trip to Kitty Hawk, leaving their mechanic, Charles Taylor, to look after their bicycle shop. They brought a tent in which to live as well as a workbench, tools, and their glider.

The Wrights' first glider was a biplane that featured a space for the operator to lie flat and manipulate the wings. Initially they flew the glider like a kite, fastened by a rope tied to its nose. Finally, Wilbur got into the craft and let Orville raise

Orville Wright working on struts of a glider at Kitty Hawk, North Carolina.

him up into the wind. When the wind picked him eight feet off the ground, however, Wilbur began to shout to be let down. Orville, thinking he was cheering to be raised higher, let out more rope. Their experiment with the first glider was thus a mixed success.

Build a wind tunnel

The Wrights returned to Kitty Hawk in 1901, with a new, larger glider, built according to information published by Langley and Lilienthal. They expected the new glider to provide enough lift to easily take them into the air in a strong wind, but this glider performed poorly. To test their data they built a small wind tunnel in their bicycle shop. It was an eight-foot-long wooden box, with a metal fan powered by a two-cylinder gas engine they had made themselves. In this box they tested over two hundred different types of wing surfaces. From these experiments they found that the information on which they had relied was not merely inaccurate—it was useless. For the next two years the Wrights compiled all their own data and used these findings to build a third glider in 1902. This craft, with a wingspan of 32 feet and a wing area of 305 square feet, broke all previous records for glider flight. They were now ready to try powered flight.

Build the Wright Flyer I biplane

At first the Wrights assumed that an automobile engine would be suitable for mounting on their glider. When they could not find an engine light enough to meet their specifications, however, they built their own. With the help of Charles Taylor, they designed a small, lightweight 4-cylinder engine that could generate 12 horsepower (a unit of power equal to 746 watts named for the power that a horse exerts in pulling). Finding no complete data on propellers, the brothers conducted their own experiments with propellers—often riding a bicycle through Dayton with a propeller attached to the front of the bike to see how it would spin in the wind.

When finished, the Wright Flyer I biplane had a wingspan of 40 feet and wing area of 510 square feet. With two engine-driven propellers, it sat on sled runners. The Wrights took this plane to Kitty Hawk in late September 1903, and for three months they worked toward perfecting it, while at the same time flying their old glider. Finally, on December 14, they laid a track on the slope of Kill Devil Hill to act as a runway. They were ready for their first powered flight. Winning a coin toss, Wilbur got to fly first. The plane rose off the ground for three seconds before falling. It had traveled 105 feet—the first true powered flight—but the brothers were disappointed in the short duration of the flight. The flight had also damaged the plane's left wing. After two days repairing the plane, it was now Orville's turn to fly.

Achieve first successful flight

The morning of December 17 was cold, with gale-force (very strong) gusts of wind. The brothers should have postponed the flight, but they had promised their father and sister Katharine they would return to Dayton in time for Christmas dinner. Orville took off into a 27-mile-per-hour wind and flew erratically for 12 seconds, traveling 120 feet before crashing into the sand. They tried again. In all, the Wright brothers made four flights that day, the longest by Wilbur—852 feet in just under a minute. They ended their experiments before noon on December 17, 1903, now considered one of the most momentous days in history. Yet at the time their achievement went unnoticed. Since numerous inventors had made bogus claims of flight in the past, newspapers refused to report the story of the Wright brothers' success, and since experts had declared powered flight to be impossible, the public was skeptical, too.

Build the world's first practical airplane

In 1904 and 1905 the Wrights improved their plane, making over 120 flights in a friend's 70-acre field outside Dayton. Their third biplane, the 1905 Flyer III, was the world's first practical airplane—it could carry two people and remain in the air for over half an hour, performing feats like sharp turns and figure eights. As their flights became smoother and more predictable—from a combination of improved design and better flying skills—they realized that their airplane could be of military significance.

Soon the Wright brothers were visited in Dayton by foreign dignitaries. The British government sent a representative from its war department, and the French dispatched a military commission. Both countries were interested in the military applications of the new invention. But the Wrights wanted the U.S. government to have exclusive use of the airplane's military potential. They wrote the War Department with a contract offer, but received an astonishing reply stating that U.S. military officials did not believe they could fly. In 1908 the U.S. Army Signal Corps accepted the Wrights' bid to make a plane

Wilbur Wright during a 1908 demonstration flight in France.

according to Army specifications for $25,000, but reportedly only because the brothers were already negotiating the sale of the airplane to foreign governments. In May of that year the Wright brothers became the first people to fly together.

Tragedy strikes

In September 1908, during army test flights at Fort Myer, Virginia, one of the Wrights' passengers was killed in a crash. Lieutenant Thomas Selfridge, who had asked the War Department to let him participate in Orville's tests, died of a fractured skull hours after a cracked propeller forced the plane down. Selfridge became the first person to die in a plane crash. Orville's left leg and several ribs were also broken in the acci-

The Langley Aerodrome

For many years the Smithsonian Institution exhibited "the first airplane capable of flight." However, it was not the Wright Flyer I, nor any other plane designed by the Wright brothers. The plane on display at the Smithsonian was the *Aerodrome,* which had been built by American astronomer and airplane pioneer Samuel P. Langley (1834–1906). Ironically, Langley's *Experiments in Aerodynamics* was the publication the Wright brothers had studied and eventually disproved. Samuel P. Langley had been the secretary of the Smithsonian Institution, and in 1898 he was given $50,000 by the U.S. War Department to develop the *Aerodrome.* An enormous craft, it was powered by a 50-horsepower engine that alone weighed 200 pounds. After spending $70,000 dollars, Langley launched the plane over the Potomac River in October 1903. It immediately fell into the water. On December 8, 1903, he tried again, but the plane sank a second time. The *Aerodrome* never did fly, and Langley died in 1906. Despite his failure, he nevertheless received more initial credit than the Wright brothers. In 1943 the Smithsonian issued an apology to Wilbur and Orville Wright.

dent. Only after news of the accident reached the press and was widely reported was the American public finally convinced that powered flight had been achieved.

At the time of Orville's crash, Wilbur was in Europe negotiating with the French government. He flew for thousands of excited French spectators and became a national hero. The French Wright Company began giving flying lessons, and by World War I (1914–18) airplanes were commonplace all over Europe. Still, the brothers did not receive recognition at home. Wilbur's last flight as a pilot was on May 21, 1910. Two years later he died of typhoid fever. Orville succeeded him as president of the Wright Company. He sold the company in 1915, and then founded the Wright Aeronautical Corporation, later becoming director of the Wright Aeronautical Laboratory in Dayton.

During their years of refining the airplane, the Wrights had relied on patents to protect them from competitors. Both in the United States and in Europe, however, they became embroiled in patent disputes. Particularly stormy was a fight between the Herring-Curtiss Company and the Wright Company. Although the courts ruled in favor of the Wrights in nearly every case, the verdicts allowed ways for competitors to work around the Wrights without violating their patents. The result was that Europeans made rapid advancements in the development of the airplane. Meanwhile, in the United States the patent fights, along with a general attitude toward flying as merely a form of entertainment, seriously impeded the progress of American aviation.

Orville develops the Whirlwind

World War I ended the patent bickering. The United States government directed the Wright Company to work with Glenn Curtiss. The potentially difficult partnership, however, turned out well for all involved. The Navy Bureau of Aeronautics also ordered Wright Aeronautical Corporation to produce an airplane that used an air-cooled engine designed by engineer Charles Lawrence. At first Orville was reluctant to comply because his own company had done research on water-cooled engines. However, he followed navy orders and produced a plane called the Whirlwind, which became an international standard almost immediately.

Wrights honored for achievements

Recognized as a scientific pioneer, Orville received honorary degrees from several institutions, including Earlham

College, the Royal Technical College of Munich, Yale and Harvard Universities, and the University of Dayton. Wilbur also received honorary degrees from Earlham and Oberlin Colleges, and the brothers were jointly awarded a medal from the French Academy of Sciences in 1909. Orville was presented the Langley Medal from the Smithsonian, the Cresson Medal from the Franklin Institute, a Cross of an Officer of the Legion of Honor, and the Distinguished Flying Cross. In 1930 he became the first recipient of the Daniel Guggenheim Medal.

In 1940 Orville was issued the first honorary pilot certificate by the new U.S. Civil Aeronautics Authority. Three years later he took the controls of an airplane for the last time, piloting the new, high-speed Lockheed C-69 Constellation, nicknamed "Connie." On January 30, 1948, he died of a heart attack. Neither Wilbur nor Orville was married, and they had no descendants. Together, however, they left a legacy that reached to the skies and beyond.

Further Reading

Crouch, Tom D. *The Bishop's Boys: A Life of the Wright Brothers,* Norton, 1989.

Haynes, Richard M., *The Wright Brothers,* Silver Burdett Press, 1991.

Heyn, Ernest V., *Fire of Genius: Inventors of the Past Century,* Anchor Press/Doubleday, 1976.

Howard, Fred, *Wilbur & Orville: A Biography of the Wright Brothers,* Knopf, 1987.

Those Inventive Americans, National Geographic Society, 1971.

Chien-Shiung Wu

Born May 31, c. 1912
Liu Ho, China

Internationally renowned nuclear physicist, lecturer, and educator Chien-Shiung Wu is best known for an experiment she conducted on beta decay in 1957. Her discovery confirmed a revolutionary prediction made a year earlier by physicists **Chen Ning Yang** and **Tsung-Dao Lee** (see combined entry) regarding the conservation of parity, the basic symmetry of nature, in reactions involving the weak force of a nucleus. A number of observers believe Wu should have shared in the 1957 Nobel Prize for physics that was awarded to Lee and Yang for their work, but she was not included among the recipients. She is, however, the first living scientist to have had an asteroid—one of thousands of orbiting rock formations between Mars and Jupiter—named for her.

Chien-Shiung Wu overcame both racism and sexism to become one of the most influential female physicists in history.

Receives Western education

Chien-Shiung Wu was born on May 31, 1912 (some sources say 1913), in Liu Ho, a small town near Shanghai,

IMPACT

Chien-Shiung Wu overcame both sexism and racism in her path to becoming a prominent nuclear physicist. Although her experiments helped confirm a prediction two male colleagues made about the conservation of parity, she was denied the 1957 Nobel Prize that was awarded to them. Nonetheless, Wu enjoyed a nearly four-decade-long research and teaching career at Columbia University that began in 1944 with her joining the Manhattan Project, the U.S. government-sponsored program to develop the atomic bomb. Throughout her career Wu has outspokenly criticized traditional scientists for their old-fashioned attitudes toward women in science.

China. Her father, Wu Zhongyi, had participated in the revolution of 1911 that toppled the Manchu dynasty. (The Manchus had ruled China for more than 250 years.) He was the founder and principal of a private girls' school, one of a very few such schools in China, a country where educating girls was not a high priority. Wu attended her father's school until the age of nine, when she was sent to the Soochow Girls' School, which had a Western curriculum (oriented toward North America and Europe). She received an excellent education at Soochow, graduating at the top of her class in 1930.

Wu went on to college at National Central University in Nanking, where she studied physics and, again, emerged as a top student. After graduating in 1934 she spent two years teaching and conducting research in the field of X-ray crystallography, a new field concerned with determining the molecular structure of crystalline materials through the deflection of X rays. She then decided to continue her studies in the United States because China did not have a physics program at the postgraduate level. With the encouragement of her family and the help of a wealthy uncle, Wu sailed for the West Coast of America in 1936.

Studies in United States

Wu's original intention was to enroll at the University of Michigan, but once she arrived in the United States she remained in the San Francisco area for a number of reasons. The University of California at Berkeley was an exciting place for a student of physics in the 1930s. Wu had also learned that the University of Michigan did not open its student union building to women, a policy she found disgraceful.

At Berkeley, Wu studied nuclear physics with the renowned particle physicist Ernest Lawrence, who invented the cyclotron, or atom smasher, and worked as an assistant to Emilio Segre, a future Nobel Prize winner. After her first year there, in 1937, Wu was recommended by the physics department for a fellowship, a grant of money that would have helped her pay for further education. She was denied the fellowship, Wu believes, because of racist attitudes toward the Chinese. She experienced yet another blow that year when Japan invaded China and she was completely cut off from her family, whom she would never see again.

Becomes expert on nuclear fission

For her thesis work, Wu studied the electromagnetic energy released when the pace of a particle passing through matter is slowed. She did another study on the results of the splitting of uranium nuclei. Wu received her Ph.D. in 1940 and stayed on at Berkeley for two years as a research assistant. During this time she became well known in the scientific community as a reliable expert on fission, the splitting apart of atoms. Nonetheless, Berkeley refused to hire her in a more formal capacity; the only conceivable reason was racial and gender-based discrimination.

Breaks gender barrier

In 1942 Wu married Chia Liu "Luke" Yuan, whom she had met shortly after moving to San Francisco, and the couple moved to the East Coast. Yuan found a job working at the Radio Corporation of America (RCA) laboratories in Princeton, New Jersey, and Wu took a position teaching at Smith College in Northampton, Massachusetts. Smith did not have a research department, however, and research was Wu's primary interest. She soon moved on to a position at Princeton University, where she taught introductory physics to naval officers. Hiring a female instructor was an unprecedented move at that time; the university had never permitted females to teach male students, especially in the sciences. Although Princeton did not have a research department, this position would at least allow Wu to live with her husband.

In 1944 Wu was recruited by the Division of War Research at Columbia University in New York City. She worked at a facility in New York on the development of sensitive radiation detectors for the Manhattan Project, a secret scientific operation that engineered the world's first atomic bombs. Staying on at Columbia after World War II, Wu finally heard from her family. They had all survived the war, and she and her husband considered returning to China. They later decided to stay in the United States because of a civil war raging in their homeland between the nationalist Chinese and the Communist forces of Mao Zedong, who would be victorious in 1949.

Begins work on parity principle

In 1956 Wu began work with Tsung-Dao Lee, who was also at Columbia, and Chen Ning Yang, who worked at the Institute for Advanced Study at Princeton. Lee and Yang were concerned with the validity of a concept of nuclear physics known as the principle of parity. This principle states that, on a nuclear level, an object and its mirror image will behave in the same way. Four years earlier a new particle called a K-meson had been discovered. Strangely, K-mesons did not seem to behave in the manner described by the principle of parity, which had been universally accepted for thirty years.

Proves Lee-Yang theory

When Lee and Yang announced that the principle of parity was flawed, they asked Wu to perform experiments to disprove the theory. She worked independently at the National Bureau of Standards, but her experiments were based on their observations. Her laboratory was underfunded and much of the equipment was below standard. Nevertheless, in January 1957 Wu's team had enough proof to say that K-mesons did indeed violate parity. The announcement took the scientific community by storm: physicists around the world were stunned by the result.

Similar work was then undertaken by physicists throughout the United States, who began finding other particles that violated this once sacred law of nature. Wu, Yang, and Lee became national celebrities, with their findings reported on the front page of the *New York Times* and in *Time* and *Newsweek* magazines. Later that year Yang and Lee won the Nobel Prize for physics. Wu reportedly was not included for her work because her experiments were not based on her own original ideas.

Receives recognition

Wu was given an endowed professorship by Columbia University in 1972. She retired in 1981 and has since traveled extensively. Through the years she has received many honors for her pioneering work. She was the first woman to receive the Comstock Award from the National Academy of Sciences, the first woman to receive the Research Corporation Award, the first woman to serve as president of the American Physical Society, and only the seventh woman selected for membership in the National Academy of Sciences. Wu has received honorary degrees from more than a dozen universities, including Harvard, Yale, and Princeton, where she was the first woman ever to receive an honorary doctorate of science degree.

Further Reading

Kass-Simon, G., and Patricia Farns, eds., *Women of Science: Righting the Record,* Indiana University Press, 1990, pp. 205–08.

McGrayne, Sharon Bertsch, *Nobel Prize Women in Science: Their Lives, Struggles, and Momentous Discoveries,* Carol Publishing Group, 1993.

McLenighan, Valjean, *Women and Science,* Raintree Publishers, 1979, pp. 42–47.

Rosalyn Sussman Yalow

Born July 19, 1921
New York, New York

If women are to "achieve status in academic medicine in proportion to their numbers.... We must match our aspirations with the guts and determination to succeed."

American medical physicist Rosalyn Sussman Yalow co-developed radioimmunoassay (RIA), a diagnostic tool that uses radioactive substances to find and measure minute amounts of hormones, viruses, enzymes, and other substances in blood or tissues, allowing doctors to diagnose a variety of diseases and medical conditions. A trailblazer for other women scientists, Yalow was only the second woman to receive the Nobel Prize for physiology or medicine, an award she won in 1977 for her groundbreaking work.

Becomes interested in physics

Yalow was born on July 19, 1921, in the Bronx, a borough of New York City, to Simon Sussman and Clara Zipper Sussman. Although neither parent had attended high school, they instilled great respect for education in their daughter, and Yalow was reading books even before she entered kindergarten. During her youth she became interested in mathematics, but at Walton High School in the Bronx, however, her interest turned to science, especially chemistry.

After graduation Yalow attended Hunter College in New York City, where she majored in physics. She entered this field in the latter part of the 1930s, a time when many new discoveries were being made in nuclear physics. Yalow credits two professors, Herbert Otis and Duane Roller, for inspiring her to choose this major. In 1939 she was further inspired after hearing American physicist **Enrico Fermi** (see entry) lecture about the discovery of nuclear fission (the splitting of the nucleus of an atom, thereby releasing energy), which had earned him the Nobel Prize the previous year.

Encounters obstacles

As Yalow prepared for graduation from Hunter College, she found that practical considerations intruded on her passion for physics. At a time when most of American society expected young women to become secretaries or teachers, she would have a difficult time becoming a scientist. She thought it was unrealistic to expect any of the top graduate schools in the country to accept her into a doctoral program or offer her the financial support men received. Fortunately, Yalow's expectations were not completely accurate. Just one month after graduating from college in 1941, she was offered a teaching assistantship in physics in the engineering college at the University of Illinois at Champaign-Urbana that allowed her to teach introductory classes as well as pursue her doctorate.

Despite this advancement in her career, Yalow still encountered obstacles as a woman in the sciences. When she entered the university, for example, she was the only woman on the engineering faculty of four hundred professors and teaching assistants. She was also the first woman in more than two decades to attend the engineering college. Furthermore, Yalow realized she had been given a space at the prestigious graduate school because of the shortage of male candidates, who were being drafted into the armed services in increasing numbers as America prepared to enter World War II (1939–45).

IMPACT

Rosalyn Sussman Yalow and long-time colleague Solomon A. Berson developed radioimmunoassay (RIA), the use of radioactive substances to find and measure minute substances in blood and tissue. A powerful tool that can quickly and precisely detect very low concentrations of hormones, vitamins, viruses, enzymes, drugs, and other substances, RIA helps scientists and medical professionals diagnose diseases and other medical conditions. Yalow was awarded the Nobel Prize in physiology or medicine in 1977 for her work with RIA; Berson had died five years earlier.

Nonetheless, Yalow excelled in graduate school. In addition to her regular course load and teaching duties she took extra undergraduate courses. She also met Aaron Yalow, a fellow student, whom she married three years later. Yalow received a master's degree in 1942 and her doctorate in 1945. She was only the second woman to obtain a Ph.D. in physics from the University of Illinois.

Becomes full-time researcher

After graduation Yalow and her husband moved to New York City, where they worked and eventually raised two children, Benjamin and Elanna. Yalow's first job was as an assistant electrical engineer at Federal Telecommunications Laboratory, a private research lab. Once again she found herself the sole woman, as there were no other female engineers at the lab. In 1946 she began teaching physics at Hunter College, where she remained for four years. In 1947 she began her long association with the Veterans Administration (VA) by becoming a consultant to the Bronx VA hospital. The VA wanted to establish research programs to explore medical uses of radioactive substances. By 1950 Yalow had equipped a radioisotope laboratory at the hospital and decided to leave teaching to devote her full attention to research. (Radioactive substances are elements that spontaneously release energetic particles by the disintegration of their atomic nuclei; isotopes are varieties of the same element; radioisotope is short for radioactive isotope.)

Begins partnership with Berson

In 1950 Yalow began a partnership with Solomon A. Berson, a physician who had just finished his residency in internal medicine at the VA hospital. They would work together until Berson's death in 1972. Yalow and Berson would discover new ways of using radioactive isotopes in the measurement of blood volume. They would also apply these methods to the study of iodine metabolism (the processes by which a particular substance is used in the living body) and the diagnosis of diseases of the thyroid (a gland that produces various hormones, or chemicals that play specific roles in the body). Within a few years they would use radioisotopes to investigate adult-onset diabetes, a disorder of carbohydrate metabolism characterized by inadequate secretion or utilization of insulin (a hormone essential to the metabolism of carbohydrates). This project eventually led them to develop the groundbreaking radioimmunoassay (RIA) technique.

Researches diabetes

In the 1950s some scientists hypothesized that in adult-onset diabetes insulin production remained normal, but a liver enzyme (a chemical produced by a living cell that sparks specific biochemical reactions) rapidly destroyed the peptide hormone, thereby preventing normal glucose metabolism. (Glucose, a sugar, is the body's primary source of energy.) This contrasted with the situation in juvenile diabetes, in which insulin production by the pancreas is too low to allow proper metabolism of glucose. Yalow and Berson wanted to test the hypothesis about adult-onset diabetes. They used insulin "labeled" with iodine-131. (That is, they attached, by a chemical reaction, the radioactive isotope of iodine to otherwise normal insulin molecules.) Yalow and Berson injected labeled insulin into diabetic and nondiabetic individuals and measured the rate at which the insulin disappeared.

To their surprise, and in contradiction to the liver enzyme hypothesis, they found that the amount of radioactively labeled insulin in the blood of diabetics was higher than that found in the control subjects, who had never received insulin injections before. As Yalow and Berson further investigated this finding, they concluded that diabetics were forming antibodies (proteins made of specialized white cells that counter foreign proteins known as antigens) to the animal insulin used to control their disease. These antibodies were binding to radiolabeled insulin, preventing it from entering cells where it was used in sugar metabolism. People who had never taken insulin before did not have these antibodies, and so the radiolabeled insulin was consumed more quickly.

Faces criticism

Yalow and Berson's proposal that animal insulin could spur antibody formation was not readily accepted by immunologists (scientists who study the body's disease-fighting system) in the mid-1950s. At the time, most immunologists did not believe that antibodies would form to molecules as small as the insulin peptide. Also, the amount of insulin antibodies was too low to be detected by conventional immunological techniques. So Yalow and Berson set out to verify these minute levels of insulin antibodies, using radiolabeled insulin as their marker. Their original report about insulin antibodies, however, was initially rejected by two journals. Finally, a compromise version was published, omitting "insulin antibody" from the title and including additional data to indicate that an antibody was involved.

The need to detect insulin antibodies at low concentration led to the development of RIA. The principle behind RIA is that a radiolabeled antigen, such as insulin, will compete with unlabeled antigen for the available binding sites on its specific antibody. As a standard, various mixtures of known amounts of labeled and unlabeled antigen are mixed with antibody. The amounts of radiation detected in each sample correspond to the amount of unlabeled antigen taking up antibody binding sites. In the unknown sample, a known amount of radiolabeled antigen is added, and the amount of radioactivity is measured again. The radiation level in the unknown sample is compared to the standard samples; the amount of unlabeled antigen in the unknown sample will be the same as the amount of unlabeled antigen found in the standard sample that yields the same amount of radioactivity. RIA has turned out to be so useful because it can quickly detect very low concentrations of

Andrew V. Schally, Biochemist, and Roger Guillemin, Endocrinologist

Polish-born American biochemist Andrew V. Schally (1926–) and French-born American endocrinologist Roger Guillemin (1924–) shared the 1977 Nobel Prize in physiology or medicine with Rosalyn Sussman Yalow. Yalow's work with radioimmunoassay (RIA) helped Schally and Guillemin isolate and analyze hormones in the brain, which greatly advanced understanding of the function of hormones and their effect on body processes. Hormones, chemical substances that play specific roles in the body, are secreted by endocrine glands. The endocrine system is a hierarchical one in which the hypothalamus, located at the base of the brain just above the pituitary gland, regulates the pituitary gland, which in turn controls other endocrine glands, including the thyroid and adrenal glands. Scientists were unsure, however, of the way in which hypothalamic regulation occurred. In the 1930s English anatomist Geoffrey W. Harris theorized that it was done by means of hormones, which are transported by the blood. The problem was that no one had yet been successful in isolating and identifying a hormone from the hypothalamus.

In the late 1950s and early 1960s, when they were both on the faculty at Baylor University Medical School in Houston, Texas, Schally and Guillemin together began an investigation to find the missing evidence, a task of extraordinary difficulty because very minute amounts of hypothalamic substances are involved. The colleagues used the new tool RIA, which enabled them to isolate and identify the chemical structure of hormones. By 1963, however, the friendly cooperation between Schally and Guillemin had given way to fierce competition. Schally took a position as director of the Endocrine and Polypeptide Laboratory at the Veterans Administration (VA) Hospital in New Orleans, Louisiana, in 1962, and Guillemin was named director of the Laboratory for Neuroendocrinology at Baylor the following year.

From the mid-1960s to the early 1970s Schally and his research team and Guillemin and his colleagues independently isolated, identified, and determined the chemical structure of three brain hormones: thyrotropin-releasing hormone (TRH, which regulates the thyroid gland), luteinizing-hormone releasing factor (LHRH, which controls reproductive functions), and growth-releasing hormone (GRH, which regulates the production of growth hormones and insulin). They also identified a factor that inhibits the release of GRH; named somatostatin, it serves multiple roles, some relating to insulin production and growth disorders. Schally's and Guillemin's discoveries proved useful in the treatment of thyroid diseases, infertility, diabetes, and a host of other hormone-deficiency diseases.

Andrew V. Schally, Rosalyn Sussman Yalow, and Roger Guillemin at a press conference in June 1977, after being notified they were to share the Nobel Prize.

hormones and other substances in blood, other biological fluids, and tissue. The principle can also be applied to binding interactions other than that between antigen and antibody, such as between a binding protein or tissue receptor site and an enzyme.

Receives awards and honors

In 1968 Yalow became a research professor at the Mt. Sinai School of Medicine, and in 1970 she was made chief of the Nuclear Medicine Service at the VA hospital. She also received a number of prestigious awards in recognition of her role in the development of RIA. In 1976 she became the first woman to be awarded the Albert Lasker Prize for Basic Medical Research, an honor that often leads to a Nobel Prize. The following year she shared the Nobel Prize in physiology or medicine with Andrew V. Schally and Roger Guillemin (see box), who used RIA to make important discoveries about brain hormones.

Acknowledges Berson

Yalow was saddened that Berson, who had died in 1972, was not able to share the Nobel Prize, which recognized the work they had done together. Earlier she had paid tribute to her collaborator by asking the VA to name the laboratory, in which the two had worked, the Solomon A. Berson Research Laboratory. Throughout her career Yalow has received numerous other awards, honorary degrees, and lectureships, including the Georg Charles de Henesy Nuclear Medicine Pioneer Award in 1986 and the Scientific Achievement Award of the American Medical Society. In 1978 she hosted a five-part dramatic series on the life of French physical chemist **Marie Curie** (see entry), aired by the Public Broadcasting Service (PBS). In 1980 she became a distinguished professor at the Albert Einstein College of Medicine at Yeshiva University in New York City. Six years later she was appointed Solomon A. Berson Distinguished Professor at Large at Mt. Sinai. She also chaired the Department of Clinical Science at Montefiore Hospital and Medical Center in the 1980s.

Further Reading

Dash, Joan, *The Triumph of Discovery: Women Who Won the Nobel Prize,* Messner, 1990.

Opfell, Olga, *The Lady Laureates: Women Who Have Won the Nobel Prize,* Scarecrow Press, 1978.

Chen Ning Yang

Born September 22, 1922
Hofei, China

Tsung-Dao Lee

Born November 24, 1926
Shanghai, China

Chen Ning Yang

Chinese-born American physicists Chen Ning Yang and Tsung-Dao Lee won the 1957 Nobel Prize for physics for their discovery of the violation of parity conservation.

Chen Ning Yang and Tsung-Dao Lee developed the revolutionary theory that the unusual behavior of the subatomic particle K-meson was a result of its violating the conservation of parity, a supposedly fixed law of nature. A few months after their theory had been announced, fellow physicist **Chien-Shiung Wu** (see entry) obtained experimental confirmation of their remarkable discovery. For their work, Lee and Yang were awarded the 1957 Nobel Prize for physics.

Yang grows up in academic family

Chen Ning Yang was born in the city of Hofei, China, on September 22, 1922, to Ke Chuan Yang, a professor of mathematics, and the former Meng Hwa Loh. In 1929 the family moved from Hofei to Peking, where Yang's father took a job with Tsinghua University. In Peking Yang attended the Chung Te middle school. By 1937 the Yangs were forced to flee the invading Japanese army and went to Kunming. When Yang finished high school he entered the National Southwest Associated University, with which Tsinghua University had merged. Majoring in physics, he earned a bachelor of science degree in 1942. He continued his studies at Tsinghua, where his father was still a professor of mathematics, and received a master's degree two years later. He then taught high school for a year before deciding to pursue a doctorate in physics.

Because doctoral programs in physics were not then available in China, Yang moved to the United States. Anxious to study with Italian nuclear physicist **Enrico Fermi** (see entry), he traveled to New York City under the impression that Fermi was working at Columbia University. When Yang heard that the noted physicist had recently taken a new post at the University of Chicago, he followed—and enrolled in the doctoral program at Chicago. One of the benefits of Yang's tenure at Chicago was the association he developed with fellow student Tsung-Dao Lee.

Lee follows professor to the U.S.

Lee was born in Shanghai, China, on November 24, 1926. He was the third of six children born to Tsing-Kong Lee, a businessman, and Ming-Chang Chang. Lee attended the Kiangsi Middle School in Kanchow and, after graduation, entered the National Chekian University in Kweichow. After the invasion of Japanese troops in China in 1945, Lee fled to the south, where he continued his studies at the National Southwest Associated University in Kunming. The following year he was presented with an unusual opportunity: one of his teachers at Kunming was the theoretical physicist Ta-You Wu; when Wu decided to return to the United States (where he had worked toward his doctorate), he invited Lee to accompany him. Lee accepted the offer but found himself in a somewhat peculiar position. He had not yet received his bachelor's degree and learned that only one American institution, the University of Chicago, would accept him for graduate study without a degree. He therefore decided to enroll at Chicago.

Yang and Lee become friends

For his doctoral study Lee selected a topic in astrophysics and worked under Fermi. In 1950 he earned a Ph.D. after completing a dissertation on the hydrogen content of white dwarfs, stars in the final stages of their evolution. While at Chicago Lee also renewed his acquaintance with Yang. (Although they had both attended Southwest University in China, Yang was a year ahead of Lee.) The pair became close friends at Chicago. They were separated in the early 1950s when Lee went to the Yerkes Astronomical Observatory at Lake Geneva, Wisconsin, and Yang joined the Institute for Advanced Studies at Princeton University. The two friends were reunited when Lee accepted an appointment at the Institute for Advanced Studies.

Begin research collaboration

Lee's later departure from Princeton for a post as assistant professor of physics at Columbia University seemed to have little effect on his collaboration with Yang. The two worked out a schedule that allowed them to continue meeting once a week, either in New York City or in Princeton. By the spring of 1956 they had begun to focus on a particularly interesting subject, a subatomic particle known as the K-meson. Discovered only a few years earlier, the K-meson puzzled physicists because it appeared to be a single particle that, when bombarded in a particle accelerator, decayed (split) in two different ways. The decay schemes were so different that physicists had become convinced that K-meson had two distinct forms, tau meson and theta meson.

The single difference between these two mesons was that one form had even parity and the other form had odd parity. The law of parity conservation defines the basic symmetry of nature, the theory that the laws of nature are not biased in any particular direction. That is, if one form has two sets of interactions that are mirror images of each other, the physical laws describing those interactions are identical. This concept is known as the conservation of parity, a concept held by physicists since 1925. By the 1950s, however, K-meson had raised questions about the validity of that law.

Confront K-meson problem

Lee and Yang were faced with vast amounts of experimental evidence suggesting that the theta and tau mesons were one and the same particle. The only contrary evidence was that the two mesons had opposite parity and, therefore, supposedly could not be identical. During an intense three-week period of work in the spring of 1956, Yang and Lee attempted to solve this puzzle. Their solution was to suggest, simply enough, that in some types of reactions parity is not conserved. The beta decay (radioactive decomposition of an atomic nucleus in which a neutron breaks apart into a proton and an electron) of the one and only K-meson was such a reaction. The duo then devised a series of experiments by which their theory could be tested. Lee and Yang announced the basis of their explanation in a paper they sent to the journal *Physical Review* in June 1956. They later titled the report "Question of Parity Conservation in Weak Interactions."

IMPACT

Chen Ning Yang and Tsung-Dao Lee called into question the validity of all conservation laws, which lie at the heart of physics. Conservation laws are rules that describe how physical quantities are preserved during physical processes or reactions. They state that a particular property—mass, energy, momentum, or electrical charge, for instance—is conserved even after a major transformation. An example would be two moving objects that strike each other. According to the laws of conservation, their total momentum after the collision must be the same as it was before the collision. During the 1950s, however, Yang and Lee conducted experiments on a nuclear particle called K-meson that raised doubts, in particular, about conservation of parity (the basic symmetry of nature). The results of their research showed that parity is not always preserved. This discovery was a revolutionary step in the field of physics because it ran contrary to the laws of conservation.

Win Nobel and other honors

Within several months of the publication of their paper, the experiments suggested by Lee and Yang were carried out by one of their colleagues, Chien-Shiung Wu, first at Columbia University and then at the National Bureau of Standards. The experiments confirmed the Yang-Lee prediction in every respect. Less than a year later, the two physicists were awarded the 1957 Nobel Prize for physics for their work.

In 1965 Yang ended his long affiliation with the Institute for Advanced Study to accept an appointment as Albert Einstein Professor of Physics and director of the Institute of Theoretical Physics at the State University of New York at Stony Brook. In 1950 he had married Chih Li Tu, who had been one of his high school students in China. They have two sons, Franklin and Gilbert, and a daughter, Eulee. In addition to the Nobel Prize, Yang received the 1957 Albert Einstein Award and the 1980 Rumford Medal of the American Academy of Arts and Sciences.

Lee was promoted to the rank of professor at Columbia in 1956. He returned to the Institute for Advanced Studies in 1960 for three years, then was appointed Enrico Fermi Professor of Physics at Columbia in 1963 and later became University Professor at Columbia. Beginning in 1981 Lee held appointments as honorary professor at a number of Chinese universities, including the University of Science and Technology and Jinan, Fudan, Quinghua, Peking, Nanjing, and Zhejiang Universities. He had married Hui-Chung Chin (also called Jeanette) in 1950, while they were both students at Chicago. The Lees have two sons, James and Stephen.

Further Reading

Crease, Robert P., and Charles C. Mann, *The Second Creation,* Macmillan, 1986, pp. 205–07.

Magill, Frank N., ed., *The Nobel Prize Winners: Physics,* Volume 2, Salem Press, 1989, pp. 707–13.

McGraw-Hill Modern Scientists and Engineers, Volume 2, McGraw-Hill, 1980, pp. 215–16.

Wasson, Tyler, ed., *Nobel Prize Winners,* H. W. Wilson, 1987, pp. 1150–52.

Vladimir Zworykin

Born July 30, 1889
Mourom, Russia

Died July 29, 1982
Princeton, New Jersey

As a researcher for Westinghouse and the Radio Corporation of America (RCA), Russian-born American scientist Vladimir Zworykin was a pioneer in the development of television, electron microscopy, and infrared vision. During his lifetime he obtained more than 120 patents on a wide variety of electronic devices, but he is best remembered as "the father of television." His research led to the development of the electronic microscope, which uses a beam of electrons instead of light to produce an enlarged image.

"After a few days I was informed, very politely, that my demonstration [of the picture tube] had been extremely interesting, but that it might be better if I spent my time on something 'a little more useful.'"

Introduced to television

Born in Mourom, Russia, on July 30, 1889, Vladimir Zworykin (pronounced ZWOR-i-kin or ZVOR-y-kyn) was the youngest of seven children. At an early age he became so fascinated by science that he would stay after school to help his teachers in the science laboratory. He was especially interested in electricity. After graduating from high school, Zworykin

IMPACT

Vladimir Zworykin's electron microscope had an enormous impact on the scientific community: by 1950 thousands of the instruments were in use all over the world. Today, the electron microscope is an invaluable research tool in nearly every branch of science. Over the years Zworykin worked on a number of unusual projects, many involving the use of computers. He suggested that computers and telephones could someday be linked for more rapid information transmission. Zworykin thus anticipated the advent of the Information Superhighway, the computer-operated system used for sending and receiving messages almost instantly.

enrolled at the University of St. Petersburg, where he majored in physics. However, his father, who wanted him to become an engineer, persuaded him to transfer to the St. Petersburg Institute of Technology.

At St. Petersburg, Zworykin's physics professor, Boris Rosing, was in charge of the institute's laboratory projects. Rosing maintained that cathode-ray tubes, with their ability to shoot a stream of charged particles, would be useful in the development of television. (A cathode-ray tube is a vacuum tube in which beams of high-speed electrons are projected on a screen by a strong electric field.) A quarter of a century earlier, in 1884, a German inventor named Paul Nipkow had patented a mechanical scanner—often called the first television—which used one spinning disk to convert a televised image into an electrical current in a transmitter and another spinning disk to create the picture in a receiver. The system was not light-sensitive enough to be practical, however, owing in part to its use of mechanical instead of electronic components.

Rosing chose Zworykin to be his only laboratory assistant in his private research on television. In 1910 the two scientists were able to demonstrate a television system that used Nipkow's mechanical scanner in the transmitter along with a cathode tube in the receiver. Like previous attempts at television, it was primitive and impractical; unlike earlier systems, however, the cathode-ray tube was more efficient and lasted longer than its mechanical-based counterpart.

Immigrates to the United States

After receiving his degree from the St. Petersburg Institute of Technology in 1912, Zworykin entered the College de France in Paris, where he studied X-ray technology under Paul

Langevin, a well-known French physicist. When World War I broke out in 1914, Zworykin returned to Russia and spent the war years as a radio officer in the radio signal corps. During the war he married Tatiana Vasilieff, and the couple later had two children. When the war ended in 1918, Zworykin left Russia and traveled widely before immigrating with his family to the United States the next year.

Invents iconoscope

In 1920 Zworykin took a job at the Westinghouse research laboratories in Pittsburgh, Pennsylvania, where he worked on the development of radio tubes and photoelectric cells (small devices whose electrical properties are modified by the action of light). Within three years he had patented the iconoscope, an electronic television transmission tube, with which he tried to reproduce the function of the human eye electronically. The iconoscope uses a lens to focus an image on a signal plate of mica, which is covered with tiny dots of photoelectric cells to simulate the function of the retina of the eye. An electronic beam, which corresponds to the optic nerve, scans the signal plate from top to bottom in parallel lines, detecting the electrical emissions. This process produces the picture. Zworykin's system, which was more sensitive than any mechanical apparatus then being explored, reduced the amount of light necessary to obtain a clear picture electronically.

Demonstrates TV system

In 1924, the year he became a U.S. citizen, Zworykin patented the kinescope, a special cathode-ray tube that could send a steady stream of electrons to the signal plate. The kinescope, or picture tube, corresponds to the function of the human brain in vision. An electronic beam is applied to an electrode grid and then adjusted through the use of electromagnetic fields. The invention of the kinescope made television as we know it possible.

Development of the Cathode-ray Tube

Cathode-ray tubes, at the center of early television technology, were first developed in the mid-1800s by Heinrich Geissler and Julius Plucker at the University of Bonn. Plucker, then a young professor, convinced Geissler, the university's scientific instrument maker, to design an apparatus for evacuating most of the air from a glass tube to aid in plucker's experiments. Geissler constructed a hand-crank mercury pump that (after a laborious session of pumping) could remove most of the air. The new tubes, known as Geissler's tubes, became very popular in scientific research.

With the improved vacuum tube Plucker was able to produce a bright, stream-like glow between the electrodes at either end of the tube—much brighter than any achieved in previous experiments. He soon found that the glow responded to a magnetic field, and that it could be moved by a powerful magnet. This discovery was monumental, for it indicated that the stream crossing the vacuum was composed of particles rather than rays. In 1869 Johann Hittorf, one of Plucker's students, placed tiny obstacles within the tube, in the path between the two electrodes. When a current was applied, the glow would be partially obscured by these obstacles, casting shadows. This further confirmed the idea that the glow was caused by a particle emission.

Probably the most important research using cathode-ray tubes was performed in 1875 by the English physicist William Crookes. Continuing Plucker's and Hittorf's experiments, Crookes designed his own vacuum tube from which the air could almost completely be removed. He installed within them tiny vanes that would turn slightly when an electric current was applied, as if they were blown by a gust of wind. These experiments proved beyond a doubt that a stream of particles traveled through the tube. And so great were Crookes's improvements to the tube, the "Crookes tube"—renamed cathode-ray tube in 1876—quickly became the standard vacuum tube for use in scientific experiments.

Zworykin demonstrated his new system to Westinghouse executives. As he later wrote, "I was terribly excited and proud. After a few days I was informed, very politely, that my demonstration had been extremely interesting, but that it might be better if I were to spend my time on something 'a little more useful.'" Although Westinghouse decided not to pursue Zworykin's research, he continued his efforts to develop the

From Cathode Ray to Kinescope

While many scientists were trying to unlock the secrets of cathode rays, others were searching for ways to apply them toward practical ends. The first such application came in 1897 in the form of German physicist Carl Ferdinand Braun's oscilloscope. This device used a cathode-ray tube to produce luminescence on a chemically treated screen. The cathode rays were allowed to pass through a narrow aperture (opening), effectively focusing them into a beam that appeared on the screen as a dot. The dot was then made to scan across the screen according to the frequency of an incoming signal. An observer viewing the oscilloscope's screen would then see a visual representation of an electrical pulse. About the same time, the final proof of the particulate nature of cathode rays was provided by the great British physicist J. J. Thomson. The cathode-ray particles came to be known as electrons, and the cathode-ray tube as an electron gun.

During the first three decades of the twentieth century, inventors continued to devise uses for cathode-ray technology. Inspired by Braun's oscilloscope, Scottish-born engineer A. A. Campbell-Swinton suggested that a cathode-ray tube could be used to project a video image upon a screen; unfortunately, the technology at the time was unable to match Campbell-Swinton's vision. As a result, it was not until 1922 that American inventor and engineer Philo T. Farnsworth used a magnet to focus a stream of electrons onto a screen, producing a crude image. Though the first of its kind, Farnsworth's invention was quickly superseded by Zworykin's kinescope, the ancestor of the modern television.

television. He received his doctorate in physics from the University of Pittsburgh in 1926.

Three years later Zworykin found a receptive audience for his invention at RCA. David Sarnoff, the company's president, hired Zworykin away from Westinghouse as associate director of the RCA electronic research laboratory in Camden, New Jersey. Around the same time, Zworykin filed his first patent for color television. Restrictions on the manufacturing of receivers, however, delayed the growth of telecommunications until the end of World War II. Zworykin eventually became vice president of laboratories at RCA; he maintained an affiliation with the company until his death.

Perfects electron microscope

Among Zworykin's other areas of research was the groundbreaking field of electron microscopy. In 1939 the Camden lab produced an electron microscope that was as large as the side of a house. Within a year the group had built a revised model that stood only six feet tall. The electron microscope was a breakthrough for research in many areas, allowing scientists to clearly see details twenty to fifty times smaller than could be detected with an optical microscope. Unlike an ordinary microscope, which uses lenses to magnify visible light, an electron microscope shoots a beam of electrons through a series of magnetic coils. The electrical impulse produces a picture in a way similar to a TV transmitter or a receiver.

During World War II Zworykin invented electron image tubes that allowed infrared "night vision" for the first time. Infrared light cannot be seen, but its waves are so long that they can pass through materials that would block visible light. Before the war was over Allied soldiers were seeing in the dark with the sniperscope and the snooperscope, two devices Zworykin had secretly developed for the war effort.

Zworykin continued working well into his seventies. During his lifetime he was recognized for his achievements in television, science, and engineering and for the application of science to medicine. In 1967 he was awarded the National Medal of Science by the National Academy of Sciences. Zworykin died the day before his ninety-third birthday, having lived long enough to witness the effects of television as a major influence on culture and events around the world. Asked to comment on American television in an interview in 1981, Zworykin replied: "Awful."

Further Reading

Those Inventive Americans, National Geographic Society, 1971.

Yost, Edna, *Modern Americans in Science and Technology,* Dodd, Mead, 1962.

Picture Credits

Jane Goodall

The photographs appearing in *Scientists: The Lives and Works of 150 Scientists* were received from the following sources:

On the cover (clockwise from top right): Luis Alvarez, Robert H. Goddard (**AP/Wide World Photos. Reproduced by permission.**); Margaret Mead (**The Bettmann Archive. Reproduced by permission.**). On the back cover (top to bottom): Edwin H. Land (**AP/Wide World Photos. Reproduced by permission.**); George Washington Carver (**The Bettmann Archive. Reproduced by permission.**).

Courtesy of the Library of Congress: pp. v, 11, 30, 66, 82, 96, 98, 102, 135, 137, 185, 188, 197, 200, 203, 206 (top and bottom), 208, 211, 224, 244, 260, 267, 275, 296, 308, 316, 339, 371, 374, 395, 399, 405, 441, 444, 487, 490, 506, 524, 527, 529, 568, 573, 599, 616, 620 (top and bottom), 648, 654, 680, 686, 693, 723, 726, 746, 759, 772, 787, 794, 803, 822, 855, 858, 905, 908, 945, 959, 990, 996, 1004, 1023; **AP/Wide World Photos. Reproduced by permission.:** pp. xv, xliii, 7, 19, 34, 72, 92, 105, 121, 124, 128, 141, 142, 233, 236, 239, 242, 254, 258, 273, 280, 282, 283, 320, 326, 333, 335, 367, 376, 384, 389, 397, 408, 415, 418, 422, 425, 432, 438, 466, 468, 474, 478, 483, 498, 502, 512, 523, 556, 575, 581, 585, 588, 642, 652, 660, 665, 674, 733, 735, 737, 738, 743,

769, 783, 797, 801, 814, 838, 845, 862, 882, 901, 917, 921, 957, 962, 969, 980, 992, 999, 1012; **The Bettmann Archive. Reproduced by permission.:** pp. xxix, 65, 144, 174, 181, 226, 458, 609, 631, 670, 728, 750, 810, 889, 923, 926, 995; **UPI/Corbis-Bettmann. Reproduced by permission.:** pp. xxxi, 112, 115, 117, 178, 251, 378, 448, 454, 515, 593; **Courtesy of Keiiti Aki:** p. 1; **Photograph by Robert T. Eplett. California Governor's Office of Emergency Services.:** p. 3; **The Granger Collection, New York. Reproduced by permission.:** pp. 27, 37, 41, 85, 168, 184, 217, 249, 392, 543, 607, 628, 651, 658, 753, 805, 825, 876, 893, 948, 987, 1017; **UPI/Bettmann. Reproduced by permission.:** pp. 48, 51, 56, 75, 95, 127, 131, 150, 155, 292, 341, 412, 463, 493, 495, 578, 603, 604, 622, 635, 639, 695, 755, 757, 836, 867, 933, 936, 951, 1010, 1025; **AT&T Bell Laboratories. Reproduced by permission.:** p. 62; **UPI/Bettmann Newsphotos. Reproduced by permission.:** pp. 80, 134, 149, 161, 780, 819, 832, 870; **Washington University Photographic Services. Reproduced by permission.:** p. 165; **From *Undersea World of Jacques Cousteau.* American Broadcasting Company. Reproduced by permission.:** p. 171; **Courtesy of Francisco Dallmeier:** p. 192; **© Zefa Germany, Stock Market. Reproduced by permission.:** p. 195; **Los Alamos National Laboratory:** p. 285; **Illustration by Robert L. Wolke. Reproduced by permission.:** p. 289; **Archive Photos, Inc. Reproduced by permission.:** p. 313; **Photri/Bikderberg. Stock Market. Reproduced by permission.:** p. 329; **Reuters/Bettmann. Reproduced by permission.:** pp. 363, 508, 565; **U.S. National Aeronautics and Space Administration (NASA):** p. 437; **J. W. Cappelens Forlag. Reproduced by permission.:** p. 451; **Courtesy of Louis Keith:** p. 549; **Photograph by Yousuf Karsh. Reproduced by permission of Woodman Camp:** p. 536; **© 1990 Peter Menzel. Reproduced by permission.:** p. 562; **Mary Evans Picture Library. Reproduced by permission.:** p. 591; **Anthony Howarth/Science Photo Library. Reproduced by permission.:** p. 596; **Courtesy of Raymond Kurzweil:** p. 668; **© Scott Camazine, The National Audubon Society Collection/Photo Researchers, Inc. Reproduced by permission.:** p. 689; **Conservation International. Reproduced by permission.:** p. 767; **Royal Institute of Technology. Reproduced by permission.:** p. 791; **Photograph by Allen Furbeck. Reproduced by permission.:** p. 827; **Courtesy of Dr. Susan Leeman.:** p. 851; **© The Alan Mason Chesney Medical Archives of The Johns Hopkins Medical Institutions. Reproduced by permission.:** p. 897; **Courtesy of New York Public Library Picture Collection:** p. 911; **Courtesy of Dr. Levi Watkins:** p. 929; **Photograph by Brent Clingman. *Time* Magazine. Reproduced by permission.:** p. 973.

Index

Italic type indicates volume numbers;
boldface type indicates
entries and their page numbers;
(ill.) indicates illustrations.

James D. Watson

A

B

C

D

E

F

G

H

I

J

K

L

M

N

O

P

Q

R

S

T

U

V

W